W9-BKS-731

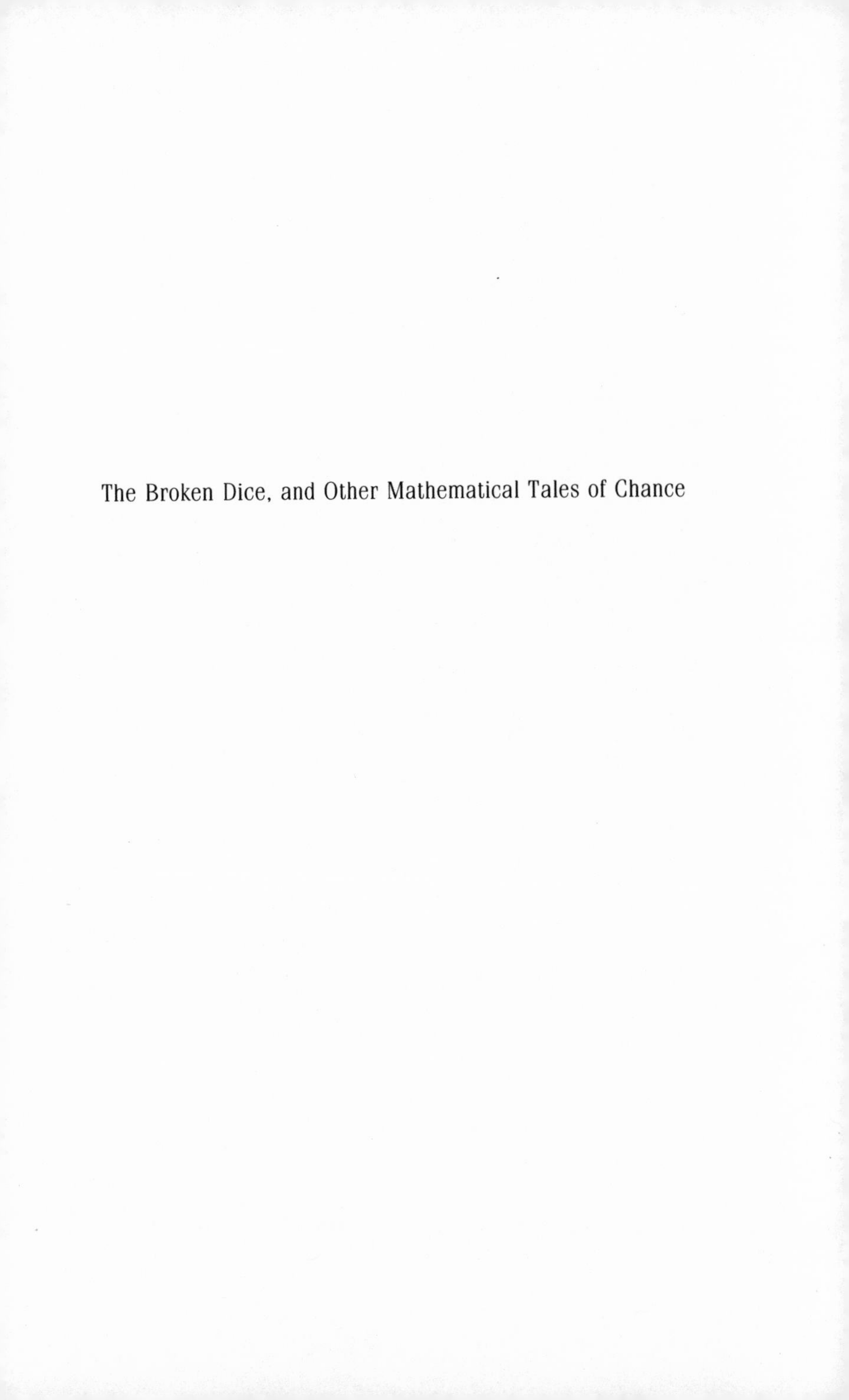

The Broken Dice, and Other Mathematical Tales of Chance

THE BROKEN DICE

and Other Mathematical Tales of Chance

IVAR EKELAND

Translated by Carol Volk

THE UNIVERSITY OF CHICAGO PRESS
Chicago and London

Ivar Ekeland is president of the Université Paris—Dauphine and founder of the Centre de récherche de mathématiques de la décision. He is author of several books including *Mathematics and the Unexpected,* also published by the University of Chicago Press.

Illustrations on chapter opening pages by John MacDonald.

Originally published as *Au hasard,* © Editions du Seuil, 1991

The University of Chicago Press, Chicago 60637
The University of Chicago Press, Ltd., London

Printed in the United States of America
02 01 00 99 98 97 96 95 94 93 1 2 3 4 5 6
ISBN (cloth): 0-226-19991-6

∞The paper used in this publication meets the minimum requirements of the American National Standard for Information Sciences—Permanence of Paper for Printed Library Materials, ANSI Z39.48-1984.

Library of Congress Cataloging-in-Publication Data

Ekeland, I. (Ivar), 1944–
[Au hasard. English]
The broken dice, and other mathematical tales of chance / Ivar Ekeland.
p. cm.
Translation of: Au hasard.
Includes bibliographical references and index.
1. Probabilities. 2. Mathematical statistics. I. Title.
QA273.E4313 1993
519.2—dc20 93-19341

Contents

Prologue

WHEREAS THE HEBREW BIBLE places the original chaos at the beginning of time, before God created the world, the Scandinavian Edda sets it during an intermediary period. It is the *ragna rök,* the twilight of the gods and the destruction of the world, between the story of the sons of Odin and the new golden age that will rise from the ruins. Here is what the Voluspo says, the vision of a prophetess, an imposing text that encompasses the entirety of the cycle:

> Axe-time, sword-time,
> shields are sundered,
> Wind-time, wolf-time,
> ere the world falls.[1]

The giants charge into battle on a ship constructed from the nails of the dead. The wolf Garm, who barks at the gates of hell, breaks his chain, the serpent Midgard emerges from the bottom of the ocean to fight the god Thor, the wolf Fenrir kills Othin, who is avenged by his son Vithar, the ash tree Yggdrasil, beneath which the three Norns decide human destinies, splits from top to bottom, the sun grows dark, the earth is swallowed up by the sea, fire destroys even the stars.

Perhaps these times are our own. Amid the chaos in which our history unfolds fragments of the ancient world and of a future golden age exist side by side, just as, under a magnifying glass, the dull sand on the beach turns into a mass of

1. From *A Pageant of Old Scandinavia,* edited by Henry Goddard Leach (New York: The American-Scandinavian Foundation, 1955).

varied, iridescent grains. This is why it seemed to me possible, and in some sense necessary, to bring together two texts separated by a millennium, a few pages from an ancient saga and fragments of a modern treatise on chance. While one text speaks of fate, magic or destiny, and the other of chance, chaos or risk, they are telling the same story. It begins in the ancient times when the same Greek word, τυχη, encompassed all these things and also meant existence. As we read this story in search of distraction or knowledge, we gradually discover that we are its characters.

Like Janus, chance has many faces, and it is the richness of these faces that I wanted to convey. I didn't want to fit this abundance of perspectives into the artificial framework of a well-pruned tree, nor impose on our thoughts the rhetorical unity of a neatly reasoned thesis. Perhaps to this end it would be useful for chance to influence the way we approach the text. Reader, this book has six chapters. Choose your die; you know what to do.

1

. . . .

Chance

Thorstein the Learned says that there was a settlement on the Island of Hísing which had alternately belonged to Norway and to Gautland. So the kings agreed between them to draw lots and throw dice for this possession. And he was to have it who threw the highest. Then the Swedish king threw two sixes and said that it was no use for King Olaf to throw. He replied, while shaking the dice in his hand, "There are two sixes still on the dice, and it is a trifling matter for God, my Lord, to have them turn up." He threw them, and two sixes turned up. Thereupon Olaf, the king of Norway, cast the dice, and one six showed on one of them, but the other split in two, so that six and one turned up; and so he took possession of the settlement.[1]

ACCORDING TO A TRADITION, a minor but a well-documented one, King Olaf Haraldsson manipulated chance on this occasion. Some say that from the beginning of a career that was to lead to canonization, he had miraculous powers such as being able to heal the sick and crippled or to conjure up help from the beyond to fight on his side. He thus had the

1. *Saint Olaf's Saga,* chap. 94. From the *Heimskringla: History of the Kings of Norway,* by Snorri Sturluson, trans. Lee M. Hollander (Austin: University of Texas Press, 1964, 1967).

power to stop the dice on the numbers he desired. For others, just as certain warriors, called *berserk,* were invested on great occasions with superhuman force that made them invulnerable, King Olaf Haraldsson was capable of supernatural ability, which allowed him to throw dice so skillfully that they naturally landed on the side he had chosen. An old chronicler even assures us that this aptitude wasn't innate, and recounts how the King had acquired it by practicing on smaller and smaller dice. Still others accuse him of out-and-out cheating. According to them, the dice were loaded—which is why the six came up with such regularity—and one of them so cleverly cracked it left no visible trace. Also according to them, Olaf Haraldsson staged the scene all the way to the surprise ending, which was a surprise for the King of Sweden and his entourage only.

It is true that every aspect of the roll of dice may be suspect: the dice themselves, the form and texture of the surface, the person throwing them. If we push the analysis to its extreme, we may even wonder what chance has to do with it at all. Neither the course of the dice nor their rebounds rely on chance; they are governed by the strict determinism of rational mechanics. Billiards is based on the same principles, and it has never been considered a game of chance. So in the final analysis, chance lies in the clumsiness, the inexperience, or the naiveté of the thrower—or in the eye of the observer.

In fact, one might perfectly imagine a civilization in which the rolling of dice would be a sport and billiards a game of chance. The rules of each would be different. The dice would be the size and weight of billiard balls, and the game would be played like the current Lyonnais game of *boule.* The player would take a few steps back, throw his die so that it would land as close as possible to a target lying ten yards away; the number that appeared on top would count toward the score, and the skillful or experienced player would modify his throw accordingly.

As for billiards, it can easily be transformed into a game of chance by simply tilting the table, outfitting it with studs that would cause the balls to rebound and swerve, and by placing the six pockets at the bottom of the table, or at other points, so that the ball would necessarily fall into one of them. Since we're not trying to favor skill, there would be a mechanical trigger and the ball would be shot up the slope by a spring that the player would pull with more or less force. This game of mechanical billiards is no less random than traditional dice. It is less inclined to this usage, since a pair of dice is pocket size and can be ready for any occasion, but in principle there is no reason why the two kings couldn't have decided the fate of the city on a contraption of this kind. A few centuries later, technological progress would change mechanical billiards into pinball and the game of chance would once again become a game of skill.

I don't share this dismal suspicion, and prefer to believe that in this circumstance, as in many others, King Olaf Haraldsson proved himself worthy of his reputation of saintliness. His recourse to chance can then be understood in several ways, notably as a divine test, a judgment of God, and it is in this way that the chronicler no doubt understood it. A modern mind might prefer to see it as a way of splitting something indivisible in two. The two kings recognized that each had an equal claim to the city. By some extraordinary fluke they had neither principles to uphold nor interests to defend, saw sharing the city as having more drawbacks than advantages, and came upon this method of deciding their claims. To give someone half of an object or to give him one chance in two of having the entire object is just about the same thing, but only the latter is possible when the object in question is indivisible. This method is familiar today in mathematical economics, where it is used to support the claim that all goods are infinitely divisible.

The method is very flexible, and lends itself to many other

purposes. For instance, if the two kings had wanted to recognize one of them as having twice the rights of the other, they could simply have had a third person roll a single die, with the city falling to one for a total of four or less, and to the other for a total of five or more. The first would thereby have had two out of three chances of winning, against one out of three for the other, which reflects the agreed-upon proportions of one to two.

Even if the two kings were in good faith, and if equity was their only concern, there would still be one problem to resolve: how to cast lots in perfect honesty, exempt from all suspicion of cheating? Is this even possible? Is chance a mere psychological attitude or social convention, or is there such a thing as pure chance, outside of all human intervention? We find this question debated in a remarkable manner in a manuscript that unfortunately has disappeared, but of which Jorge Luis Borges sent me a copy he had made at the Vatican archives. According to him, the manuscript dates from the years 1240–50, and was no doubt placed in Olaf Haraldsson's file during his canonization hearings. It was written by a certain Brother Edvin, from the Franciscan monastery of Tautra in Norway, of whom nothing else is known to us.

Upon reading the text, we are hardly surprised at the author's obscurity. Brother Edvin's audacity is too reminiscent of his contemporary Roger Bacon's for the latter's condemnation not to have affected him. Perhaps they even followed the teachings of Robert Grosseteste together at Oxford, and perhaps Edvin, like his contemporary, finished his days incarcerated. If Brother Edvin composed other writings, they undoubtedly did not survive the condemnation of 1277; the one at hand must have either been saved from destruction by an oversight or, perhaps, been protected as evidence for conviction.

Brother Edvin begins by putting an end to the accusations of cheating leveled against King Olaf. This rumor appeared belatedly, long after the disappearance of the last eye-witness, whereas a tradition was formed among the Christian people immediately upon the death of the King affirming his saintliness in all circumstances. This tradition, which is still alive and well today, is bolstered by numerous miraculous signs, not the least of which is the episode mentioned, a manifestation of the efficacious support Our Lord accorded his disciple. The splitting of the die can only be interpreted as a miracle: making thirteen out of twelve. The miracle of the loaves and fishes comes immediately to mind, but we needn't evoke such grandiose examples for a matter as trivial as the fate of a city. In any case, this was the interpretation of the witnesses, as manifested by the attitude of the King of Sweden and his entourage, who were not reported to have contested the result, as they would certainly have done had there been the slightest doubt as to its miraculous nature.

In the end, unlikely as it seems, even if Olaf Haraldsson had manipulated fate on this occasion, in no way would it call into question his saintliness. No one would question the saintliness of the patriarch Jacob. Yet, as recounted in Chapter 30 of Genesis, during the settling of a contract with his father-in-law, Laban, Jacob behaved in a manner which in an ordinary man would have been called cheating, but which, in a man of God, simply helped to carry out the wishes of the Almighty. For the satisfaction of the editor and the edification of the reader, let's turn to the sacred text:

Laban said, "What shall I give you?" Jacob said, "You shall not give me anything; if you will do this for me, I will again feed your flock and keep it; let me pass through all your flock today, removing from it every speckled and spotted sheep and every black lamb, and the spotted and speckled among the goats; and such shall be my wages. So my honesty will answer for me later, when you come to look into my wages with you. Every one that is not speckled and spotted among the goats and black among the lambs, if found with me, shall be counted as stolen." Laban said, "Good! Let it be as you have said." But that day

Laban removed the he-goats that were striped and spotted, and all the she-goats that were speckled and spotted, every one that had white on it, and every lamb that was black, and put them in charge of his sons; and he set a distance of three days' journey between himself and Jacob; and Jacob fed the rest of Laban's flock.

Then Jacob took fresh rods of poplar and almond and plane, and peeled white streaks in them, exposing the white of the rods. He set the rods which he had peeled in front of the flocks in the runnels, that is, the watering troughs, where the flocks came to drink. And since they bred when they came to drink, the flocks bred in front of the rods and so the flocks brought forth striped, speckled, and spotted. And Jacob separated the lambs, and set the faces of the flocks toward the striped and all the black in the flock of Laban; and he put his own droves apart, and did not put them with Laban's flock. Whenever the stronger of the flock were breeding Jacob laid the rods in the runnels before the eyes of the flock, that they might breed among the rods, but for the feebler of the flock he did not lay them there; so the feebler were Laban's and the stronger Jacob's. (Gen. 30:31–42)

As to the question of whether it is appropriate to roll dice to decide the ownership of a city, it rests on the highest authority possible, since the Lord's tunic was raffled off, as attested unanimously by the four evangelists, John being the most explicit on this point:

When the soldiers had crucified Jesus they took his garments and made four parts, one for each soldier; also his tunic. But the tunic was without seam, woven from top to bottom; so they said to one another, "Let us not tear it, but cast lots for it to see whose it shall be." This was to fulfill the scripture: "Diviserunt sibi vestimenta mea, et super vestem meam miserunt sortem."[2] (John 19:23–24)

It should be noted that the lottery winner isn't specified, and that, in the Scriptures as in tradition, we lose track of the

2. "They divide my garments among them, and for my raiment they cast lots" (Ps.: 22:18).

precious object from this moment on. It was therefore a matter of affirming a principle rather than relating a particular destiny, the principle being that indivisible things must not be divided. Very early on, tradition recognized the seamless tunic as an image of Mother Church, the unity of which had to be preserved from heretics, schismatics, and other followers of the devil. But the same principle applies to things of lesser importance, like a city, which is indivisible by nature, and Olaf Haraldsson's gesture is therefore entirely legitimate.

We could just as easily invoke other texts of lesser importance. "The lot is cast into the lap, but the decision is wholly from the Lord" (Prov. 16:33). Matthias is chosen by lot to complete the twelve apostles (Acts 1:26), as is Zachary to enter into the sanctuary (Luke 1:9). It is by lot, *Ourim* or *Toummim*, that the Almighty designates the guilty, Jonathan (1 Sam. 14: 37–43), Jonah (Jon. 1: 1–10), Achan (Josh. 7: 10–23), and by which Saul is designated King of Israel (1 Sam. 10: 20–24). Following the expression of Augustine (*Enarrationes in Psalmos* [Ps 30:16, enarr. 2, serm. 2]), "*Sors non est aliquid mali, sed res, in humana dubitatione, divinam indicans voluntatem.*[3]

Until now, Brother Edvin's theory is irreproachable, challenging neither the rules of logic nor orthodox belief. But from this moment on, he gets carried away by his subject and strays from the path of caution.

After firmly establishing, in his first section, that the well-conducted casting of lots is only the manifestation of divine will, Brother Edvin poses the practical problem of knowing how to eliminate the ever-present threat of human interference. The second part of the manuscript consists of a long review of the various games of chance known at the time: cards, dice, and lotteries, and shows how a clever and ill-disposed manipulator can hinder the outcome, and therefore prevent the divine will from being expressed. He concludes

3. "Lots are not bad in and of themselves, but indicate the Divine will when man is in doubt."

that the use of material instruments will always leave room for doubt as to the regularity of the operation, causing the results to be open to contestation.

In the end he proposes a very original solution. On important occasions, like the one that united Olaf Haraldsson and the King of Sweden, when it is important to everyone that the casting of lots be incontestable, each of the partners should choose a number in secret and write it on a sealed roll of parchment. On the appointed day, the two kings and their representatives hand the parchments to an arbiter, a learned and pious man, assisted by clerks who are able to perform calculations. The arbiter breaks the seals and reads the two numbers aloud; the clerks add them, divide the total by six, and announce the remainder. There are six possible remainders:

1 2 3 4 5 0

which are equivalent to the six possible results of a throw of a die:

1 2 3 4 5 6

and the number announced is considered the result of the lottery. Thus, if one chooses 17 and the other 3,051, the total will be 3,068 and the remainder 2. The arbiter will announce this number. According to our table, this result corresponds to a roll of two on a die, but Brother Edvin's method offers the advantage of being impervious to skill or ill will. It's chance, as pure as a thirteenth-century clerk can create it.

Brother Edvin then engages in some extremely interesting mathematical considerations. He points out that, if we were to multiply instead of adding, the procedure could be easily manipulated: if one player chooses a multiple of six, the result of the drawing will be 0 no matter what number his opponent picks. Indeed, we would then be dealing with the product of two numbers rather than their sum, and if one of the factors is divisible by six, its product will be so as well. His second observation is that all that matters for the final result is the remainder of the two chosen numbers after they are divided by six. Thus, we saw that if one chooses 17 and the other 3,051, the result will be 2. If we replace 17 by 5 and 3,051 by 3, which is what remains after dividing each by six, we find 5 + 3 = 8, which once again gives a result of 2. Brother Edvin concludes from this that it is pointless for the players to choose large numbers: in no way are they restricting the possibilities if they limit themselves to numbers between one and six. Still, he observes, the greater the range of possible numbers the greater the illusion the players have of increasing their chances, and it is therefore a good idea for them to have the choice.

The last part of the manuscript discusses various ways of improving the method. It extends to the case of three players or more, and Brother Edvin points out that even if only two players have an interest in the outcome, there might be some advantage to introducing a third. Thus, when important interests are at stake—like the fate of a city—one can ask the Holy Father to send a third number from Rome, to be unsealed along with two others. The three will be added together, and whatever remains after dividing the total by six will be the result. Thus, if we add 442 to the two numbers 17 and 3,051, the total becomes 3,510 and the result 0, which corresponds to one die landing on 6. The introduction of the third player, and his independence from the two others, serves to further guarantee total impartiality.

Finally, Brother Edvin mentions that occasionally one might have cause to play alone; he wonders how to consult fate without the intervention of another player. He admits to not having arrived at a satisfactory solution, but provisionally proposes the following method. The player chooses a four-digit number and squares it. He thereby obtains a seven- or eight-digit number of which he eliminates the two last and the first one or two digits, so as to once again obtain a four-digit number. He repeats this operation four times, divides the final number by 6 and obtains a remainder. If he starts with 8,653, he squares this number (74,874,409) and keeps the four middle digits (8,744). Repeating the operation, he successively obtains:

8,653 8,744 4,575 9,306 6,016

Dividing the last number by 6, he obtains a remainder of 4, which equals a throw of the die of 4.

The advantage of this method is that, aside from certain exceptions, it is impossible to predict the final result without running through the calculations—which, as Brother Edvin points out, is not within everyone's abilities, since it requires either familiarity with the abacus or knowledge of "Indian Calculus," as detailed in Leonardo Fibonacci's *Liber Aba.*[4] Thus, the remainder when the number originally chosen is divided by six is 1, whereas the final result after squaring four times is 4; one can hardly predict this beforehand. Brother Edvin also indicates that the method can be made even safer by increasing either the number of digits retained (six instead of four, for example, starting with a six-digit number), or the number of operations, provided of course that the rules are decided from the start and that the player doesn't change them in the middle of the game.

But Brother Edvin is equally prompt to point out the flaw of his system, a flaw that resides in the very existence of exceptions. You don't have to be a genius to predict that if you start with 0000, the successive numbers will all be 0000, and the final result will therefore be 0. The appearance of zeros may disturb the game in an even more insidious manner. If you start with 1,001 for example, you get 0,200 after the first operation, then successively 0,400, 1,600, 5,600, 3,600, 9,600, 1,600, after which a loop is formed, that is, the cycle 1,600, 5,600, 3,600, 9,600, 1,600 repeats itself indefinitely. It

4. Greek and Roman notation did not lend itself to these operations, and the slightest calculation using symbols was a feat.

therefore becomes perfectly possible to predict the result of the fourth, the seventeenth or the millionth round, and thus to cheat with oneself.

Brother Edvin proposes a few remedies, notably that the first number chosen have four different digits and no zeros. But he is too smart not to realize that, in addition to these obvious exceptions, more subtle ones may exist. Perhaps these exceptions are just an indication of a more profound regularity that is invisible to our untrained eyes but that may be brought to light by a more inspired analysis. How can we be sure that there isn't a hidden formula, not just in special cases like 1,001, but for all numbers? The fact that we are unable to find such a formula is no guarantee that it doesn't exist. And if it exists, our game is no longer a game of chance. Whoever finds it can either keep it a secret and make a fortune by taking bets, or reveal it and destroy the results of all these efforts.

It is on this melancholy note, perhaps a premonition, that the manuscript ends. It is easy to imagine the trouble it must have caused its author. Brother Edvin is manipulating numbers as if they were objects, without concern for their occult meaning. What good can come of such a reductive approach? Everyone knows, for example, that the number of the Beast is 666 (Revelation 13–16). What would happen if this number appeared in the course of the calculations? Wouldn't it be an explicit sign of the devil at work? How can we think that the result would be untainted? By abandoning the traditional interpretations, Brother Edvin exposes himself to accusations of practicing divination, and there is reason to fear that it was held against him.

After a gap of several centuries, the problems raised by Brother Edvin and his moral scruples became current again when it was necessary to program randomness into computers. It was no longer a matter of disinterested speculation or of judging saintly acts, but of constructing a thermonuclear weapon, the hydrogen bomb, the prototype of which was detonated in 1952. This success was the culmination of an unprecedented scientific and technological effort, initiated by the United States after Einstein's famous letter to President Roosevelt, and marked notably by the bombings of Hiroshima and Nagasaki.

The atomic-bomb computations had been performed on noisy electromechanical devices, chomping away on punch cards. ENIAC, the first electronic computer—a megalith 30 meters long, 3 meters high, and 90 centimeters wide, with 18,000 double triode vacuum tubes in a system with 500,000 solder joints—was ready to function at the end of 1947. During its long gestation period, the decision was made to use it to simulate the behavior of neutrons in fissile material—a crucial step in the development of a thermonuclear bomb. The experts who had already constructed the A-bomb, notably Enrico Fermi and John von Neumann, had looked into the problem and concluded that it could be treated only by statistical methods. At every instant, a neutron has a certain probability of entering into collision, and each collision has a certain probability of being a simple diffusion of the incoming neutron, or a fission producing several new neutrons. Each individual trajectory is therefore the result of a game played following complex rules, but with known probabilities.

The idea was to let the machine play a large number of these games, making decisions haphazardly, following the indicated probabilities, and of statistically studying the results. This was the Monte Carlo method. While waiting for the ENIAC to go into operation, Fermi had invented a small chariot, immediately baptized FERMIAC, that shuttled about on a cross section of the reactor to simulate an imaginary neutron's course, choosing the direction of the displacement and the distance to the next collision at random each time. Other adjustments were made, in particular depending on the material through which the neutron was passing. This little machine was put in storage when ENIAC arrived, which was itself outmoded after 1952 by the installation of MANIAC at Los Alamos. On all these machines, the Monte Carlo method gave excellent results. It remains today one of the main mathematical tools used in physics.

The Monte Carlo method, since that is its name, is simple and versatile. To repeat one of Stanislaw Ulam's examples, let's suppose I want to know the probability of winning a game of solitaire. It depends, naturally, on the original order of the deck of cards. And there are:

$$52! = 52 \times 51 \times 50 \times 49 \times 48 \times \ldots \times 3 \times 2 \times 1,$$

different ways of distributing 52 cards. This number (the factorial of 52—that's what the exclamation point stands for) is enormous—written with 70 digits; so large that it would be impossible to systematically examine all the possible distributions to count the winning hands. As a result, the probability sought:

$$\frac{\text{number of games won}}{\text{number of possible distributions}}$$

cannot be calculated exactly. On the other hand, we can give an empirical estimate by playing only a small number of distributions—a few hundred, or a few thousand—provided that the cards are carefully shuffled after each round. In other words, we estimate the unknown probability by randomly and independently drawing a certain number of distributions from among the 52! and by looking at the proportion of games won during such a sampling. It is easy to program a computer to play a game of solitaire, and thus to conclude in a few milliseconds if a certain distribution will win or not.

In addition, the computer must be taught to shuffle the cards, that is to say to choose a distribution from among the

52! possible, each with the same $1/52!$ probability, and independent of the preceding choices. In practice, we will represent the cards by numbers, and the computer must therefore draw a first number from 1 to 52 at random, then a second from among the 51 remaining, then a third from among the 50 remaining, and so on until they're all chosen. But how can the cards be dealt in an equiprobable fashion, and independent of the shuffles?

In bridge, this is done by shuffling the cards after each round. This is less simple than it seems. A cheater knows how to shuffle the cards so as to distribute the best hand to himself, and magicians know how to retrieve a card that has been slid into the middle of the deck. Shuffling well isn't enough; you also have to shuffle a long time—at least seven times according to recent studies. If these precautions are taken, the players won't try to remember the card order from the previous round to make conjectures about the current one: they will consider the two rounds to be independent. Yet once the cards are distributed and before the players begin declaring, they will establish an idea of the global distribution of the cards based on their own hands. This idea will be based, not on their memories of the preceding draw (the ace of spades beat the king, I noticed that the king followed the ace in the pile, and since I have the ace of spades in my hand, the king must be to my left), but on the assumption that the remaining cards are uniformly distributed (I have four spades, nine remain for three players, it is likely that each one has three). This is what we mean when we say that the hands are equiprobable.

Whether we shuffle cards or roll dice, chance is only a result of our human lack of deftness: we don't have enough control to immobilize a die at will or to individually direct the cards in a deck. The comparison is an important one nonetheless, and highlights the limits of this method of creating chance—it doesn't matter who rolls the dice, but we wouldn't let just anyone shuffle the cards. If you're too clumsy you wouldn't be able to do it, if you're too skillful you arouse suspicion. How can it be done with a computer, which is entirely incapable of being clumsy, and whose conduct is never uncertain? This is basically the problem that Brother Edvin raises, and though our technical methods have progressed enormously, the same cannot be said for our analytical means.

The method of squaring four-digit numbers and keeping only the middle four digits at each step, as depicted by Brother Edvin, was the first one used (today it bears the name of von Neumann), but it was quickly noticed that, choose as you will the first number, after a certain number of draws, the number 2,100 will appear without fail, then 4,100, then 8,100, then 6,100, then 2,100 again, and that from there on they will repeat themselves indefinitely:

2,100, 4,100, 8,100, 6,100, 2,100.

The exceptions are the very same ones Brother Edvin had mentioned, their only flaw being that they turn into a loop (different from this one) more quickly! A moment of reflection quickly convinces us. The computer takes a four-digit number, squares it a certain number of times, each time retaining only the four middle digits, and the final result is a new four-digit number, reputed to be drawn by chance. To draw several numbers at random, we iterate the procedure, that is to say that we use the last number obtained to get the next number. Thus each number depends entirely and exclusively on the preceding one. If the first selection is 8,653, the second 8,744, and the third 4,575, we can state with certainty that each time 8,653 comes up, 8,744 and 4,575 will follow. One of the fundamental characteristics of chance, the independence of successive selections, is negated in this simulation, and this is why cycles appear.

Think of a roulette croupier with little imagination who controls his wheel with a pedal and wants to hamper the random selection of numbers by making them appear in a precise order. He establishes as random a list as possible, and holds to it. After 7, 35 will always come up, after 35, 13, after 13, 22, and so on, drawing a different number each time so as to maintain an element of surprise. But since there are only 37 numbers on the roulette wheel, counting the zero, after a maximum of 37 choices he will be forced to pick a number that has already appeared—7 for example. From there, he repeats his previous choices, 35, 13, 22, successively, and on and on. How long will it take the players to catch on to his trick?

Just as this croupier will produce a cycle of 37 numbers at most, the system of successive square numbers, or any other determinist method, will end up in a cycle of 10,000 at most, 10,000 being the total number of four-digit numbers. We can look for temporary remedies, working with five-digit numbers, for example, which will open the possibility of having a cycle of 100,000, but this fundamental limitation subsists. We can also mask reality, in the manner of Brother Edvin, by making the final result the remainder of division by 6. This allows us to pretend we are choosing a number between 0 and 5, whereas in fact we are choosing a number between 0 and 9,999. Thus, the successive selections:

6,016   1,922   6,940   1,636,

would read:

4   2   4   4,

and we will have given the illusion of chance, since a four can be followed by a 2 or by a 4, as if the selections were independent. Unfortunately, the first 4, which represents 6,016, is not the same as the second, which represents 6,940. The machine works on 6,016 and 6,940 and shows 4 and 4, like an illusionist who fools the audience with mirrors.

In our day, the method of using successive square numbers for drawing random numbers has been almost entirely abandoned. We prefer to use arithmetic generators, described by mathematical formulas such as:

$$X_n = aX_{n-1} + c \text{ modulo } M,$$

which means that the *nth* draw, denoted by X_n, is obtained by taking the $(n - 1)$th result, denoted by X_{n-1}, multiplying it by a, adding c, and dividing it by M. The remainder of this division is X_n, the *nth* random number, the result of the *nth* draw. The next random number, X_{n+1}, is obtained from X_n in the same way X_n is obtained from X_{n-1}. The integers a, c, and M are characteristics of the generator; they are chosen once and for all.

These arithmetic generators suffer from the same flaws as the square-numbers method. Each draw is completely determined by the preceding one, and cycles of a maximum size of M will therefore occur. In practice, we can easily choose $M = 2^{23}$, which makes division by M easy for machines that work on a binary system, or $M = 2^{31} - 1$, which makes division by M only slightly more difficult and has the advantage of being a prime number. For such large numbers, the size of the cycle is so large it cannot be observed in practice: the number 2^{30} is in the billions. We may therefore feel we have simulated chance to a satisfactory degree.

Unfortunately, this is not the case. The question of cycles is only the first problem; there are many others along the way. In fact, the notion of chance breaks down into a multitude of properties, so diverse they sometimes seem contradictory. Thus, though we have spoken of the independence of successively selected numbers, we haven't mentioned their distribution. What we would like is for the distribution to be uniform, that is to say, with regard to the squaring method for example, that all the numbers from 0 to 9,999 appear with the same frequency. Yet we know that every series of numbers ends up stabilizing into a cycle, in general:

2,100, 4,100, 8,100, 6,100, 2,100,

these four numbers therefore appearing with a frequency of 1/4, the others not appearing at all (0 frequency). So uniform distribution will exist only during a transitional period, before the computer has had time to reach the cycle and during which we may hope that the successive selections will be nearly uniformly distributed between 0 and 9,999.

We can adjust the arithmetic generator's parameters, a and c, so that their cycles will be long. We can even organize it so that every point from 0 to $M - 1$ will be included in one cycle. The distribution would then be uniform, insofar as the M numbers from 0 to $M - 1$ each appear once per cycle, and therefore all have the same frequency: $1/M$. But what the computer is really doing in this case is simply churning out the first M numbers in an order different from the natural order: are we justified in calling this type of operation chance? Here again, chance is in the eye of the beholder; it is our inability to take in a billion numbers or more, combined with our ignorance of the rule which the computer is using to arrange them, that makes their succession seem random. A more astute observer might be able to detect hidden regularities in their distribution that will alert him to the fact that he is not dealing with chance.

Here is an example of such a situation. Suppose that we want to pick a point along the interval (0,1) following a uniform distribution. Let's first decide on the precision we'll be working with, 32 bits for example. This means that the computer will only consider numbers whose binary representation consists of a maximum of 32 signs. Essentially this means replacing the interval (0,1) with a network of $M = 2^{32}$ equidistant points between 0 and 1. After that, we will choose a point at random using an arithmetic generator:

$$X_{n+1} = aX_n + c \quad \text{modulo } M.$$

We can adjust the constants a and c so that there will be only one cycle, M in size; the M points of the interval (0,1) will each appear once per cycle. This means that they all have the same frequency, $1/M$, and that we have created a random selection that is uniform for the interval (0,1).

But what happens if we want to select a random point in a square, still using a uniform distribution? Say that the side of the square equals 1; each point in the square is therefore represented by two numbers, x and y, each between 0 and 1; one represents its horizontal coordinate (the x axis), the other its vertical coordinate (the y axis). If as before we replace the interval (0,1) with M equidistant points, we obtain M possibilities for the horizontal coordinate x and M possibilities for the vertical coordinate Y, which corresponds to $M \times M = M^2$ possibilities for the point (x,y). We therefore obtain M^2 uniform points distributed in the square. To draw one of these points at random, all we have to do is choose each of its two coordinates, x and y, successively. If these draws are independent and uniformly distributed, we can obtain any point on the square, each with the same frequency $1/M^2$.

But the two draws must be independent, and this is where we may find fault with the arithmetic generator. Let's use it to select the point's two axes, first the horizontal, x, then the vertical, y. These two coordinates are supposed to be independent. In fact we know that they're not, since the generator obtains each number from the preceding one, and therefore y from x, using the formula:

$$y = ax + c \quad \text{modulo } M.$$

Since there are M possibilities for x, each one entirely determining y, there will only be M possibilities for the duo (x,y), instead of the M^2 that would be possible by pure chance. The arithmetic generator has access to only M of the M^2 points in the square; a ratio of $M/M^2 = 1/M$. That is to say that the ma-

An example of an arithmetic generator:

This is a rudimentary model, which allows us to "randomly" choose an integer between 0 and 9. If we iterate beginning with X = 0, we obtain successively:

$X_0 = 0$	$X_6 = 8$
$X_1 = 3$	$X_7 = 1$
$X_2 = 6$	$X_8 = 4$
$X_3 = 9$	$X_9 = 7$
$X_4 = 2$	$X_{10} = 0 = X_0$
$X_5 = 5.$	$X_{11} = 3 = X_1$

that is, a complete cycle, passing through all integers between 0 and 9. This boils down to writing them in an order different from the natural order. The computer remembers the last number furnished, and each time it is called upon gives the next number on the list.

This generator can be used to "randomly" draw a point from the interval [0,1]. To do this, we replace the interval by 10 equidistant points:

$$0 = \frac{0}{9}, \frac{1}{9}, \frac{2}{9}, \frac{3}{9}, \frac{4}{9}, \frac{5}{9}, \frac{6}{9}, \frac{7}{9}, \frac{8}{9}, \frac{9}{9} = 1$$

which we can represent geometrically by:

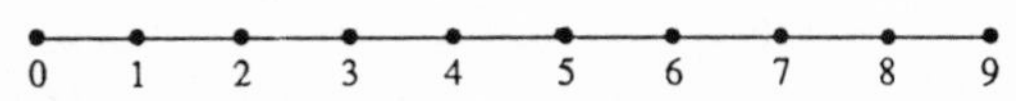

Arranging the points as we arranged the integers, we obtain a rule of succession that we can represent symbolically (each point is linked to its successor by an arrow):

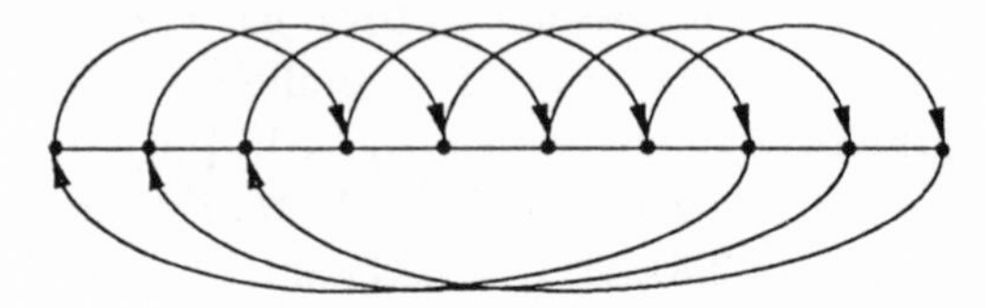

In ten iterations we hit every point on the interval. Here again, the computer remembers the last position, and gives us

$X_{n+1} = X_n + 3$ **modulo 10**

the next number each time we ask it to randomly choose a new position. Naturally this procedure is not random, but retains many of the properties of a succession of independent and equidistributed draws—enough of them, anyway, to fool an inattentive observer.

But the shortcomings of this arithmetic generator become immediately apparent when we try to use it to draw random points on a square. The square $[0,1] \times [0,1]$ is replaced by $10 \times 10 = 100$ points:

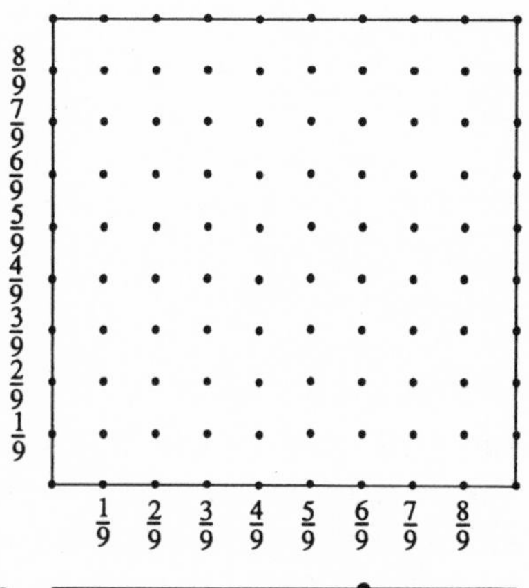

and we "randomly" draw the two coordinates of each point one after the next. Thus, if the horizontal coordinate is 0, the vertical coordinate has to be the successor of $^0/_9$, namely, $^3/_9$. This produces a total of 10 possibilities for 100 points:

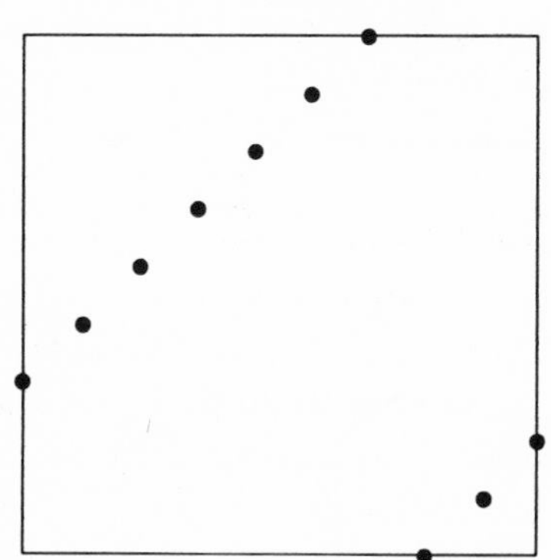

This time, there's no doubt about it: the points are not equidistributed. The draws are not independent.

Arithmetic generators commonly in use are equipped with more subtle subdivisions and far longer cycles ($M \approx 2^{32}$). Nonetheless, the existence of these cycles can lead to some unpleasant surprises.

jority of points are inaccessible. In addition, there is no guarantee that the M possible points are uniformly distributed in the square. Our cycle of M points weaves its way through an ensemble of M^2 points. This cycle could easily be concentrated in certain areas of the square while others might be completely empty.

We do have a means of detecting whether a succession of numbers, evenly distributed between 0 and 1, is truly random: we can group them in pairs, chart points in the square, and examine them to see whether they are evenly distributed in the square. If they are, we can't complain: they pass the test. But if they aren't, we can confirm that the numbers under consideration are linked, that is to say that each one depends on those preceding it. They don't pass the test.

This is what we call a test of independence. An arithmetic generator can successfully pass this test: it is possible to chose the coefficient a and the integer M (the value of c has little effect) so that the points obtained seem uniformly distributed within the square. But there are many other tests for independence, and a generator that fools one may be detected by another. For instance, we may group the numbers by clusters of three instead of two, thereby getting points in the three-dimensional cube instead of the two-dimensional square. A succession of numbers which shows up as evenly distributed in the square may very well show up as irregularly distributed in the cube, thereby passing the two-dimensional test for randomness but failing the three-dimensional one. A truly random sequence, of course, should pass both. In fact, each test has its favorites, that is, it recognizes certain generators and not others. In the spirit of the old adage, the computer can fool one test all the time and all the tests some of the time, but it can't fool all the tests all the time.

The problem here is that an arithmetic generator doesn't have the inexhaustible richness of chance. If $X_1, X_2, \ldots, X_n$ are independent and equidistributed, we can pull out every other number, regroup them by clusters of two or reverse their order, and the result will still be independent and equidistributed. More complicated transformations such as X_1^2, X_2^2, X_n^2 (each number is squared) will yield yet more independent numbers that will follow a distribution that is no longer uniform but which can be calculated. On the other hand, if an arithmetic generator creates an equidistributed series X_1, $X_2, \ldots, X_n$, there is no guarantee that by regrouping the terms in clusters of two we will obtain an equidistributed series on the square or that if we square the numbers they will be distributed according to the desired law. We can write it into the specifications at the generator's conception, but what will happen if they are regrouped into clusters of three or if they are cubed—X_1^3, X_2^3, X_n^3? And what will happen if we feel like changing the order? Will the new series behave like a random series? We can program the generator to meet these requirements and others, but we always run the risk that once in use it will encounter a test that hadn't been foreseen a priori, and will give itself away. Whereas chance can hurdle any obstacles set in its path.

At this stage in our speculation, as we realize that the most powerful computers are incapable of reproducing the properties we attribute spontaneously to chance, the gnawing suspicion returns. Does chance really exist, or are we victims of an illusion? Might it be an essentially mathematical notion, an idealization of reality, the way the geometrician's sketch represents geometry's idealized flat and infinite straight line?

The practitioner no longer poses the question. The physicist who wants to calculate an integral by the Monte Carlo method isn't looking to imitate chance in general. The series of numbers provided by the computer is of interest to him only insofar as it is the fastest way of obtaining a good approximation of the number in question. He will generally be working in a space of specific dimensions—say N—and for him what matters is that when the numbers selected are regrouped in N-size clusters the points should be equidistributed in this space. But he has no interest in the outcome of tests for independence or equidistribution that have nothing to do with his problem. In fact, of the infinite properties that a series of independent random draws contains, the user will retain only those that interest him directly, and will build his arithmetic generator accordingly; the final result sometimes loses even a semblance of randomness.

There is another possible approach, which consists of accepting our inability to construct truly random series, and of looking for chance where it is found, that is, in nature. This is why certain generators of "random" numbers combine an arithmetic procedure, similar to those we described, with the input of the clock—an instrument always present at the heart of a computer. At certain stages in the calculation, the computer simply checks the time, and brings this to bear on the operation. We can, for example, instruct our arithmetic generator to use as its point of departure X_0, the time that has passed since 5:00 P.M. on July 14, 1990, expressed in seconds. A natural random number thereby reinforces an artificial one. We can push this recourse to the outside world much farther. To construct an arithmetic generator adapted to one's needs is a difficult task, and using the available software is risky. One may therefore be tempted to use the tables of random numbers that were obtained experimentally rather than by mathematical algorithms. The first tables were of demographic origin, but the makers quickly turned to physical devices. In 1955, for instance, the Rand Corporation published a list of one million numbers extracted from electronic background noise. Unfortunately, wiring defects were noticed several years later that tainted the results and compromised the independence of successively drawn numbers. This shows that it is barely any easier to obtain randomness with a physical mechanism than with a mathematical algorithm, especially when one wants to produce very long series.

We are therefore at an impasse, like the two kings in the story told by Torstein Frode, when the sixes appeared three times in a row and everyone began feeling uneasy. This is when chance intervenes, reversing our view of the situation, shattering the limited framework of our predictions by creating something truly new, like Alexander cutting the Gordian knot. Reality takes off its mask, the indivisible element breaks in two, the number that could only be a one or a six becomes a seven. Nature laughs at us and we become ridiculous, outsiders on the edge of a world that defies our gaze.

This is the way physicists felt when the scientific revolution began with the discovery of radioactivity, when so many things that had been taken for granted turned out to be wrong. The atom had barely been established as the primary component of matter, the fundamental unity whose very name implied indivisibility, when it turned out to be composed of electrons gravitating around a nucleus, which in turn soon exploded into neutrons and protons. But quickly enough we realized we hadn't yet attained the ultimate reality. As we destroyed protons and neutrons in increasingly powerful accelerators, we noticed the appearance of other "elementary" particles among the remains: the pion, the lambda, the sigma, the rho—currently there are more than four hundred of them. In the 1970s, quantum chromodynamics introduced even more elementary components: quarks. Certain particles (the baryons) decompose into three quarks, others (mesons) into one quark and an antiquark. As I am writing, we know of five different quarks, and we suspect the existence of a sixth. In addition, there are six other particles with spin 1/2 (leptons), which do not decompose into quarks, notably electrons and neutrinos. Looking at particles with spin 1, we find twelve others: photons, gluons (eight), and Ws (three). We are therefore up to twenty-six elementary components, the diverse combinations of which should allow us to account for all varieties of known particles: for the moment it seems that the picture is complete. That is, of course, until a more ambitious theory comes along and upsets the picture. Physicists have not yet abandoned their dream of the "great unification," a theory that would encompass quantum mechanics and general relativity, and each step in this direction brings profound changes to our vision of the world.

Each time we think we have found the ultimate component of reality, the fundamental building block of the universe, it breaks into pieces like King Olaf's die. For Platonists, this inevitably evokes Parmenides' eighth hypothesis. More prosaically, it's a hunt in which the prey always gets away. The physicist runs after the ultimate reality like Coyote trying to catch Roadrunner. We sympathize with his frustration and the care he takes to devise ever more ingenious traps to ensnare his nemesis. We know that his efforts will be in vain and that he will fall flat on his face again, but he is so determined and comes so close each time that we cannot help but be awed at the cartoonist's cruelty and ingenuity. With this experience as our model, we are better equipped to appreciate the way nature evades us, and particularly the way it hides behind chance.

Once we reach the scale of the molecule, we are in the realm of quantum mechanics, which of course makes some incursions into the macroscopic domain, bringing phenomena like superfluidity or superconductivity into the realm of our perception. The theory is like a diptych, the first half of which is purely deterministic, depicting the evolution of isolated physical systems. Each of them is represented by a state vector within a space of infinite dimensions—Hilbert space. The evolution itself is purely deterministic; it is governed by a differential equation, Schrödinger's equation, which is situated within a Hilbert space instead of within the usual three-dimensional space we live in. If we wanted to be perfectly rigorous, we would be forced to envisage only one state vector, that of the entire universe, but, as in classical physics, we agree to make approximations and to consider that certain subsystems are isolated, at least momentarily, and therefore have their own state vectors: particles, atoms, or molecules.

The other half of the diptych is purely probabilistic. It depicts the operations of measurement. To measure the physical size, position or speed, energy or date, is to transfer the system from the first half of the diptych into the second. The result of this transfer will be random. More precisely, each measurement can give a variety of different results, each of which has a probability that can be computed from the state vector. The actual result of the experiment, the one the instrument will record, will be drawn randomly by nature from the possible results according to the computed probabilities.

Quantum mechanics would therefore be purely deterministic if there were no observer. By demanding information, by taking measurements, we are the ones who perturb the evolution of the system and introduce an element of randomness. It is impossible to predict the result of an isolated measurement. The most the theory does is allow us to calculate a priori all the possible results, and the probabilities of each. This doesn't mean that quantum mechanics isn't precise or doesn't allow us to make certain predictions. But these predictions are statistical in nature, based on a large number of measurements or on macroscopic phenomena—which can be very precise. As an example, the electron's magnetic moment has an experimental value of 1.001 159 652 21 (with a margin of error of about 4 on the final digit), and a theoretical value (therefore calculated a priori) of 1.001 159 652 46 (with a margin of error five times greater). This represents a precision of $4 \cdot 10^9$, which is to say a gap on the order of one millimeter for four thousand kilometers.

What a strange theory! Let's look, for example, at the movement of a photon, an elementary particle of light. We emit the photon at a point *E* (the emitter) and retrieve it at point *R* (the receptor); between the two is a screen with two holes, *A* and *B*. Optical geometry tells us that in order to observe the light at point *R*, it must be aligned, either *E* with *A*, or *E* with *B*. Quantum mechanics, on the other hand, teaches us that no matter what the position of the holes, and in particular if there is no alignment, a photon emitted at point *E* always has a certain probability of striking point *R*. This is effectively what we observe if the holes at *A* and *B* are small enough. This of course contradicts basic intuition, which would tell us that the photon is a corpuscle that is propagated in a straight line, but this isn't even the most surprising aspect of the situation. Still according to quantum mechanics, the photon that we encounter at *R* has a certain probability of having passed through point *A*, and a certain probability of having passed through point *B*—without our ever being able to determine with any certainty which path it indeed followed. Yet as we are talking about an elementary particle, which by its very nature is indivisible, it seems logical to assert that it has to go through *A* or *B*—not both. Yet again, interference fringes can be detected at point *R* that can only be accounted for by the fact that the photon passed by *both A* and *B*. If we insist on pinning down the photon at *A* or *B*, by using two detectors—one at *A* and one at *B*—to observe its passage, we find that the photon indeed passes by A or B (only one of the detectors is activated), but then that the interference fringe disappears!

Introducing an additional measurement instrument therefore changes the phenomenon. By trying to locate the photon at *A* or at *B*, we force the system into a state that is foreign to its spontaneous evolution, and thereby introduce an element of randomness. Let's say, for example, that arriving at a measurement requires such a complex interaction between the system and the observer that parameters foreign to each—but represented in the wave function of the universe—play a crucial role, and that the result can only be understood statistically. This is just a hypothesis, maybe even a metaphor. Only one thing is certain: in quantum mechanics, to measure means to draw at random.

The question that immediately comes to mind is: "Okay, but who's drawing?" It's not the observer, probably not the particle either. There is one possible answer, but not everyone will like it. We may simply not ask the question, like Niels Bohr. But if we do ask it, like Einstein, and if we reason that God doesn't roll dice, we find ourselves at an impasse—unless we show that this random-draw process itself is only an illusion.

Hence Einstein and his disciples' determined efforts to prove the existence of "hidden variables" in quantum mechanics. The thesis he defended until his death is that we have access to only some of the variables that determine the state of a quantum system. If we could observe them all, we could predict the evolution of the system—at least in the short term—and the result of any measurement. But some are invisible to us, and it is this ignorance that creates the illusion of randomness, just as an observer, sitting under a transparent table on which a game of cards is being played, would never see more than the back of the cards and would be unable to understand the evolution of the game.

Despite Einstein's innermost conviction, the theory of hidden variables has never had the slightest confirmation. On a conceptual level, von Neumann and many others tried to show that it was incompatible with the basic laws of quantum mechanics. While they weren't able to rule it out entirely, they did establish that such a theory would have to possess properties at least as paradoxical as those of quantum mechanics. On an experimental level, Einstein, Podolski, and Rosen indicated a way of approaching the question which over the years, thanks notably to Bell's discovery of certain inequalities that would be violated if hidden variables really did exist, led to feasible computations. They all turned out negative. We are therefore stuck with the idea that the randomness or chance that is a part of quantum mechanics cannot be reduced to an underlying determinism. As for macroscopic determinism, which reigns at our level, it can be reduced to quantum randomness thanks to statistical laws that embrace an immense number of particles. It is randomness or chance, therefore, that seems to be the basic fact, nature's ultimate message.

Finally, the quest for true randomness has ended at some enormous machine that spies on elementary particles. Perhaps the future will succeed in domesticating quantum randomness and in dispensing it in miniature devices that we will find in school calculators and slot machines. All of us, calculators and players, will have access to the very source of chance, pure and steady. But this domesticated chance will no longer surprise us; we will expect a choice between several familiar outcomes. Instead of looking to new technologies, we will look to history for the shock of the unexpected, for the joy of seeing the horizon suddenly expand and for the kind of trepidation that new discoveries once held in store—all the emotions we experience when the die breaks open and the seven appears.

2

• • • • • •

Fate

ing Olaf then proceeded to the town of Tunsberg and held an assembly there at which he proclaimed that all those who were known to be guilty of practicing magic and sorcery or who were warlocks must leave the country. Then the king had a search made in that neighborhood for such persons, and summoned them to his presence. Among those who came was a man called Eyvind Kelda. He was the grandson of Rognvald Rettilbeini, a son of Harald Fairhair. Eyvind was a sorcerer and exceedingly skilled in wizardry. King Olaf had all these people put in one room and entertained well with strong drink. And when they were drunk he had the house fired, and it burned down with all those inside, except that Eyvind Kelda escaped through the louver and got away. And when he had got a long ways he encountered people who intended to journey to the king, and he bade them tell the king that Eyvind Kelda had escaped and that he would never after get into the clutches of King Olaf and that he would behave as he had done before in practicing his sorcery. And when these men came before King Olaf they told him what Eyvind had bidden them. The king was greatly vexed that Eyvind was not dead.

Toward spring King Olaf sailed out along the fjord, visiting his large estates, and sent messengers about all the Vík District that he would collect troops in summer and journey north with them. Then he proceeded north [west] to Agthir. And as Lent approached, he sailed to Rogaland and by Easter Eve arrived at Ogvaldsness on the Island of Kormt. There, the Easter repast was prepared for him. He had with him close to three hundred men. The same night Eyvind Kelda approached the island with a warship fully manned with warlocks only and other kinds of sorcerers. Eyvind left his ship together with his crew and began to exert his spells. He made such a cover of darkness with fog that the king and his people should not be able to see them. But when they came close to the building on Ogvaldsness it became bright day. And then matters turned out differently from what Eyvind had intended: then the same darkness he had produced with his magic enveloped him and his followers so that they could not see any more with their eyes than with the back of their heads and went about in circles. But the king's watchman saw them but did not know what band it was, and told the king. He and all his followers arose and put on their clothes. And when he saw Eyvind and his band he ordered his men to arm themselves and go up to them to find out who they were. And when the king's men recognized Eyvind they captured him and his crew and led them to the king. Then Eyvind told him about his doings; whereupon the king had them taken out to skerries which were covered with water at high tide and bound them there. Thus Eyvind and all his companions lost their lives. That place was thereafter called Skrattasker [Sorcerers' Skerries].[1]

ALL NORWEGIANS know this gruesome story. In the classical edition of the *Heimskringla,* the one my grandfather passed on to me, it is accompanied by a splendid illustration by Eilif Peterssen. We see Eyvind's profile above the waves, his entire life concentrated in his eyes; in the distance the water, sky, and earth form an indistinct horizon, as all around him his companions are being swallowed up by the sea. I still wonder today what he was thinking at that moment, Prometheus tied

1. *Saga of Olaf Trygvesson,* chaps. 62–63. From the *Heimskringla.*

to his rock, waiting for the ocean to swell one last time. Did his mind rise above a thirst for vengeance and pity for his companions to grand metaphysical speculations? Was he asking eternal questions? Why are things as they are? Why this instead of nothing?

According to Wittgenstein's immortal definition, *die Welt is alles, was der Fall ist,*[2] the world is everything we find to be the case, everything that happens. Our very first experience is that things are just so, stubbornly so, with no explanation to give for being what they are; in philosophical terms, it is the experience of contingency. This rock is here, it is real; so are these ropes that are binding me and the rising tide. All this constitutes my universe, and it is here that I must find my place. The child, eager to discover the world, asks one question after another; wearied by age, we meditate and respond:

> *Die Ros' ist ohn' Warum; sie blühet, weil sie blühet*
> *Sie acht'nicht ihrer selbst, fragt nicht, ob man sie siehet.*[3]

But is contingency complete or is there room for meaning? Must we be content to merely note the facts, or should we look for reasons? Do events follow one another randomly, or does the world function according to certain rules that we can reveal and make use of? We often don't like the way things are, and some people go so far as to give their lives to change them; the quest for meaning must therefore be part of human existence. But before reaching this level of debate, let's look at our origins, at the appearance of life on earth, at the evolutionary tree on which *Homo sapiens* may be just a transitory branch. It provides us with a first response: each ecological

2. *Tractatus logico-philosophicus, I.*
3. The rose does have no why; it blossoms without reason,
Forgetful of itself, oblivious to our vision.
(Angelus Silesius: *The Cherubinic Wanderer,* trans. Maria Shrady [New York: Paulist Press, 1986].)

niche is an oasis of regularity, within which the various species have developed according to a rigorous logic. Since the first biped created fire, humankind has turned the resources of its brain to seeking the regularities of the world and has used them to its profit to supplant other forms of life. The search for meaning has always been undertaken with an action in mind, in general an aggressive one. It will therefore be a difficult exercise for human reason to rise above the immediate necessities of action, and it is not even clear that it is adapted to such an effort.

Let's try our luck anyway and attempt to picture the members of *Homo sapiens* in context. Their sensory organs transmit a continuous influx of information to the brain for processing. Through a mechanism that escapes us, the brain analyzes this information, finds the prototypical situation, and extrapolates in time. Thus, in the primitive savannah, the retina receives a packet of photons which the brain classifies into various shapes, among which is recognized the shape of an edible plant. The brain will order the extraction of the root and activate the complex procedure of grinding, washing, and cooking it, at the end of which it will finally become an edible food.

This type of action relies on an almost infinite number of rules noted in the past and projected into the future, and on the ability to recognize in the present the situations to which the rules apply. All these rules constitute the meaning, at least an operational meaning, that we give the world. To say that the world is meaningless would mean that we cannot detect any rules, that we don't understand the past and that we cannot predict the future; it would mean total contingency, which seems incompatible with even the precarious existence of an individual conscience. To say that the world is meaningful would signify, if this meaning were perfectly understood, that the past and the future are open before us like a book. The truth is somewhere in between, that is to say that we can figure out partial, limited meanings that allow us to function in certain circumstances, whereas in others we are left powerless.

Let's illustrate the situation by a very simple example. Let's imagine that sensory information is presented to the brain in the form of a continual series of bits, that is to say 0s or 1s:

00101000110110 . . . ,

This may seem farfetched, but it is precisely the way we communicate with computers. To feed a TV picture into a computer, for instance, we would have to translate it into this kind of binary information. This can be done in several ways. We can, for instance, decompose the picture into individual dots, as is done with newspaper photos, and specify for each dot a shade of gray (if the picture is black and white) or a color. The dots can be numbered, as can the tints, and these numbers can be written in binary code, that is, using only 0s and 1s, and the end result, which will be fed into the computer, is indeed a (long) string of 0s and 1s.

Let us now imagine that, instead of our mortal and limited brain, there is a demon, endowed with an unlimited lifespan and intelligence. The great physicist Maxwell already used such a demon to illustrate the way certain physical laws actually worked. Here, the demon receives only the information that arrives by the sensory channel, each instant adding a new bit (that is, a 0 or a 1) to the string. Since he's a demon, he lives forever. So at the end of time he has accumulated an infinite series of 0s and 1s, and we ask him then if the world makes sense.

This way of posing the question evades several major problems, notably those of perception (how does the brain classify the information it receives?) and of action (which relates to perception, since the brain doesn't passively receive information but commands actions as a consequence, notably to confirm information received and to look for more). But at least it is a question we can answer.

Of course, if the infinite string that our demon observes over the course of time happens to be constant, composed solely of 0s:

0000000 . . . ,

or solely of 1s:

11111111 . . . ,

he will immediately say yes to our question.

Similarly if the series is periodic, alternating regularly between 0 and 1:

01010101 . . . ,

or:

001001001 . . . ,

then we are dealing with a particularly simple and transparent universe. There would be three physical laws in this last case:

after 1 comes 0
after (1,0) comes 0
after (0,0) comes 1

If the series is neither constant nor periodic, our demon will try to reduce it to a few simple rules. These rules may be evident, as for the series:

0100110001110000111100000 . . . ,

(a 0 followed by a 1, then two 0s followed by two 1s, then three 0s followed by three 1s, etc.), or:

0100011011000001010011100101 . . .

The latter was introduced into mathematical literature by

D. G. Champernowne.[4] It is formed by writing all the possible combinations of 0 and 1, one after the other, first those with one digit (0 and 1), then those with two digits, then three, and so on. All these combinations are written in lexicographic order, that is to say that among combinations of the same length we write the one that contains only 0s first, 00 . . . 0, introducing the 1s at the end and finishing with 11 . . . 1. We therefore begin with 0 followed by a 1 (combinations with one digit), we then write 00, 01, 10, 11 successively (combinations with two digits), continuing with the three-, four-, five-, and six-digit combinations, and so on.

Any given combination of 0 and 1 will appear in the series not once but an infinite number of times. For instance, 10 appears as the third combination of two digits. But it appears before this, as the meeting of 1 and of the first digit of 00, and after this, notably in the three-digit combinations 010, 100, and 110. It will also appear in the combinations with four, five, six digits and more, for as far as we go into the series. In fact, if we consider two consecutive bits to be a message (there are therefore four possible messages: 00, 01, 10, and 11), the frequency of the message 10 in this infinite series is $1/4$.

The Champernowne series therefore presents all the classical characteristics of successive rounds of heads or tails: each message appears with a frequency that depends on its length. There are two one-bit messages, 0 and 1, which appear with a frequency of $1/2$. There are eight three-bit messages, from 000 to 111, which appear each with a frequency of $1/8$. The complete works of Shakespeare, once transcribed into bits, constitute a message of an approximate length of 1,000,000,000, a billion bits, which, like all messages of this length, will appear with a frequency of $1/2^{1,000,000,000} \times 999,999,999 \times \ldots \ldots \times 3 \times 2)$. This frequency is so weak

4. *Journal of the London Mathematical Society* 8 (1933): 254–60.

that the actual event has no chance of being observed between two big bangs, but our demon has eternity before him, so he can read and reread the Tragedies.

In short, the Champernowne series imitates chance so well that we might be fooled if its rule of formation weren't so obvious. But all we have to do is complicate this rule to confuse the observer. For example, let's take a simple mechanism that constructs a number, X_{n+1}, from the preceding X_n, following the formula:

$$X_{n+1} = 1 - \mu X_n^2$$

Let's assign the parameter μ the value 1.5, choose an initial value of $X_0 = 0$, and generate the infinite series X_1, X_2, X_3, . . . with the help of the preceding formula. The first twelve terms are:

$$\begin{aligned}
X_1 &= 1 \\
X_2 &= -0.5 \\
X_3 &= 0.625 \\
X_4 &= 0.4140625 \\
X_5 &= 0.7428283691 \\
X_6 &= 0.172309021 \\
X_7 &= 0.9554644019 \\
X_8 &= -0.369368335 \\
X_9 &= 0.7953505496 \\
X_{10} &= 0.05112625479 \\
X_{11} &= 0.9960791591 \\
X_{12} &= -0.4882605368
\end{aligned}$$

Then let's replace X_n by 0 if $X_n < 0$ and by 1 if $0 < X_n$. We obtain a series of 0s and 1s that begins:

$$1,0,1,1,1,1,1,0,1,1,1,0$$

and if we continue this exercise to obtain a few supplementary terms:

1,1,1,1,1,0,1,1,1,0,1,0,1,1,1,1,1,0,1,1,1,0,1,0,
1,1,1,1,1,0

This time, the formulating rule is no longer apparent, and a human observer might be tempted to assume that he is faced with a series of random numbers, independently drawn. By taking the mean of the first numbers observed, he would obtain empirical values for the respective frequencies of 0 and 1 that would serve as a basis for his statistical calculations—here, for example, we find empirical frequencies of $^{10}/_{42}$ for the 0 and $^{32}/_{42}$ for the 1 in the first forty-two terms. Thanks to the hypothesis of independence, the statistician can also evaluate the theoretical frequencies of different messages, and compare them to the empirical frequencies. Thus if the selections were independent and if the frequencies of $^{10}/_{42}$ and $^{32}/_{42}$ corresponded to reality, the pair 01 should have a frequency of $^{10}/_{42} \times {}^{32}/_{42} = {}^{320}/_{1764}$ (about 0.18), the same as the pair 10, whereas the pair 00 should have a far weaker frequency, $^{10}/_{42} \times {}^{10}/_{42} = {}^{100}/_{1,764}$, or about 0.06.

It happens that our little machine, like the Champernowne series, perfectly imitates independent draws so well that the frequencies calculated will be close to the frequencies observed; our statistician will therefore take comfort in the idea—the mistaken idea—that he is dealing with chance. He has found meaning in the world, but probabilistic meaning. Because the true determinism escapes him, he thinks that on the microscopic level phenomena are dictated by chance, and that physical laws can only be statistical.

But our demon doesn't suffer from the intellectual limitations of human physicists—we might even say he is omniscient—and he will therefore be able to trace the series:

1,0,1,1,1,1,1,0,1,1,1,0,1,1,1,1,1,0,1,1,1,0,1,0,1,1,1,1,1,0,1,1,
1,0,1,0,1,1,1,1,1,0

to its constitutive law $X_{n+1} = 1 - \mu X_n^2$. For him, the world is purely deterministic. The parameter $\mu = 1/2$ and the initial condition $X_0 = 0$ determine the entire evolution. The final refuge of contingency, the only question that still remains unanswered, is: why this world instead of another, that is, in the final analysis, why these values instead of others? Once they are known, the universe no longer holds any surprises. The diversity of the terms in the series is only superficial, their infinite number is just an illusion. All you need to know are two numbers, μ and X_0, to know the rest. If, for example, my demon wants to transmit the first forty-two terms in the series to a colleague, then it would be far more economical to send the message "$\mu = 1/2, X_0 = 0$" than to transcribe the list that appears above this paragraph, given, of course, that the recipient knows the formula $X_{n+1} = 1 - \mu X_n^2$.

Let's look at it another way and try to imagine what a world without meaning, a purely contingent world, would be like. In our model, it can be reduced to an infinite series of 0s and 1s which the demon himself cannot figure out. It simply exists, without any deterministic mechanism to generate it. The first consequence of which is that it is impossible to write it down! For if I give the first thirty terms:

$$000110110111110001010100010011\ldots,$$

though I put three dots to indicate that the series continues, no one can guess that the thirty-first term is a 0. This piece of information can only be seen and noted, and not inferred from the preceding terms. The only way to communicate the thirty-second term is to write it down, so that it would take a million 0s or 1s to communicate the first million terms in the series, another million for the second. The entire series would be incommunicable unless we had *The Book of Sand,*[5] the famous work containing an infinite number of pages that Borges misplaced on the shelves of the Argentine National Li-

5. Jorge Luis Borges, *El libro de arena* (Buenos Aires: Emecé Editores SA, 1975).

brary. Luckily, our demon has found it and is busy jotting down every term in the series.

We are therefore approaching a definition of contingency. An infinite series will be called contingent if it cannot be defined more economically than by transcribing it completely. It may not be contingent on our scale as poor humans. For if I take a contingent series and add 10^{1000} zeros[6] in front of it, it will still be contingent, and the demon, who can view the entire series, knows this, but I will die before I ever realize it, along with the entire human race. There are just too many 0s to be inspected before the first 1 appears, and to our limited view the series appears to consist only of 0s.

Here we must clarify our thoughts somewhat. Let's imagine compartments, each of which contains a 0 or a 1. Given a message of N bits (the first N terms in the series for example), we can always communicate it by simply copying it, which requires N compartments. But there are ways of being more clever. We can first indicate to our correspondent how many 0s and how many 1s there are. Each of these digits, written in binary form, occupies at most $\log N$ compartments[7] and we therefore used $2 \log N$ compartments (or $\log N$ to indicate the number of 0s and as many to indicate the number of 1s). On the other hand, we have considerably reduced the uncertainty, for with n_0 0s and n_1 1s we can create only

$$\frac{(n_0 + n_1)!}{n_0!\, n_1!}$$

6. That is to say an enormous number, easily larger than the number of known particles in the universe.

7. The notation $\log N$ designates the logarithm N in binary form, that is, a number that appears on all hand-held calculators, and which turns out to grow much more slowly than N. *Grosso modo,* it is the number of digits necessary to write N in binary form. As an example, the logarithm of 1 is 0, the logarithm of 2 is 1, the logarithm of $1/2$ is -1 and the logarithm of 2^n is n. The logarithm of a number less than 1 is negative, which explains the appearance of the $-$ sign in the entropy formula.

distinct messages. In the second part of the transmission, we will have to indicate only the order number of the desired message, which occupies a number of compartments equal to the logarithm of the preceding number. Once all these calculations have been made, we find that, for sufficiently large N, we can transmit any message of N bits by using, at most,

$$- \mathrm{N} \left(\frac{n_0}{N} \log \frac{n_0}{N} + \frac{n_1}{N} \log \frac{n_1}{N} \right)$$

compartments. Let's remember that n_0 is the number of 0s and n_1 the number of 1s.

The quantity in parenthesis is famous: it is the entropy of the message in question. This definition was introduced by C. E. Shannon, the founder of the theory of information.[8] It is a positive number (the negative sign in front of the formula compensates for the sign of the negative logarithms), representing the quantity of information transmitted by each message to a correspondent who knows n_0 and n_1. We can rewrite it by bringing in the empirical frequencies of 0 and 1, $p_0 = n_0/N$ and $p_1 = n_1/N$:

$$-(p_0 \log p_0 + \mathrm{p}_1 \log p_1).$$

In the case of equal frequencies, $p_0 = \mathrm{p}_1 = 1/2$, entropy equals 1, and we find that despite our clever tactics we still need N compartments to encode the message. This is exactly what a simple transcription would have required, so we haven't gained anything. On the other hand, if the frequencies are different from one another, $p_0 = 1/3$ and $p_1 = 2/3$ for example, entropy equals 0.9182958343, and so we can transmit the

8. C. E. Shannon, "A Mathematical Theory of Communication," *Bell System Technical Journal* 27 (1948): 379–423 and 623–56. See also the book by A. I. Khinchin, *Mathematical Foundations of Information Theory* (Dover: New York, 1957).

message by using only $0.9183 \times N$ compartments instead of N, an 8% improvement. In the extreme case, if the message contains only 0s (or only 1s), entropy is zero. There is only one message containing only 0s, so if the correspondent knows that $p_1 = 0$ we needn't tell him anything more: he will know the content of the message before receiving it. In this sense, the actual message will bring him no additional information. Inversely, if the correspondent is informed that $p_1 = 1/2$, this leaves him the choice between an enormous number of possible messages, and knowing which one will be transmitted represents important information.

Now we are in a position to clarify this idea of a totally contingent universe, a universe without rules to help us make predictions: it must be impossible to compact the messages we receive from it. To communicate the first N terms in the series to our demon therefore requires N bits. In mathematical terms, entropy has to equal 1, or more exactly it must approach 1 when the message becomes sufficiently long.

Here a problem arises. A world without rules will inevitably display great regularities at times. It is certain, for example, that somewhere within our series there will be one thousand consecutive zeros. For if not, we would have put our finger on one of the laws of the series, that "there cannot be one thousand consecutive 0s," from which we could draw an unfailing rule of prediction: "After nine hundred and ninety-nine 0s comes a 1." It is also certain that this succession of one thousand 0s will reoccur an infinite number of times. If not—if for example this event could occur only seventy-seven times—the same law, "After nine hundred and ninety-nine 0s comes a 1," would become true after the seventy-seventh succession of one thousand 0s had passed, and we would have found another rule for making predictions.

But a sequence as organized as one thousand 0s in a row can be communicated far more economically than by recopying the 0s one after the other. All you have to do is say: "Count one thousand 0s from this point." Because of these kinds of regularities, we can inevitably transmit the first N terms of the series with less than N compartments; the entropy therefore will not exactly equal 1, but will be slightly inferior.

To take another example, if we align typesetter's symbols instead of 0s and 1s, if for example we imagine a monkey pounding a typewriter keyboard, the result will be a written text, perhaps an infinite text if the process continues eternally. Most of the time, the text will have no meaning, but we can imagine that from time to time, among the "gfwsavk" and the "jmuuzxnmlkj," a recognizable word will appear by chance, most often a short one, such as "is," but sometimes a longer one, such as "island." Only rarely will we find a complete sentence, but our demon can wait forever. He has time to wait for the first sentence of *War and Peace,* time to wait for the first paragraph, time to wait for the first chapter, time to wait for the entire book, time to wait for two copies, time to wait for four thousand five hundred ninety consecutive copies. During all this time, other works are written before his eyes, this book for example, including the chapters to come which I haven't even written yet, its translation into English and other languages, and all the books that have ever been written, and all the books that no one wrote. And if the demon isn't familiar with *War and Peace* or *Hamlet,* he will have occasion to read them a first time, which will allow him to recognize the other copies as they appear. He can therefore shorten the message considerably by replacing the text *in extenso* by the note "in this place, an additional copy of such and such a work." Each insertion of this sort diminishes the entropy, whereas each succession of incoherent symbols brings it closer to 1.

Here then is our final definition: a series is contingent if the entropy of the first N terms comes ever closer to 1 as N grows. There will be for instance a number N_0 such that the entropy of any message of the length $N \geq N_0$ will lie between

0.99 and 1, another number $N_1 \geq N_0$ such that the entropy of any message of length $N \geq N$ will lie between 0.999 and 1, and so on. In other words, to communicate the first N terms of the series requires a little less than N compartments, or, in the language of computer theory, N bits. We owe this definition to the great Russian mathematician A. N. Kolmogorov (1903–87), the founder of the theory of probabilities. Among his many claims to fame is his introduction of entropy as an analytical tool in this type of question. Kolmogorov's original definition of contingency presented great difficulties on the level of pure logic, which were finally dealt with by the Swedish mathematician P. Martin-Löf.

A contingent series, according to Kolmogorov and Martin-Löf, is thus the formalization of a world in which the only rule is that there are no rules. Note that for the moment it is a purely logical construction, in which chance, in the sense of the calculation of probabilities, has no place. To highlight this point I have preferred to refer to these series as "contingent" rather than "random," as Kolmogorov or Martin-Löf did. Such series are clearly random, in the sense that there is no way of guessing a term in the series from its antecedents, nor even of condensing the information contained in its first N terms; but they are not obtained by randomly drawing 0s and 1s, as in the classical model of probability theory.

The miracle is that the two points of view converge. According to one of Kolmogorov's fundamental intuitions, the contingent series that we have described is also random according to probability theory. In fact, we can show that if a series is contingent according to Kolmogorov and Martin-Löf, the frequency of 0s and 1s in any given sampling is always close to $1/2$.

Let's look at the implications of this property. Not only can we say that there are about as many 0s as 1s in the first N terms of the series, and that the respective proportions of 0s and 1s are closer to $1/2$ the larger N becomes. We can also say that whatever the method of sampling or extraction chosen, the sample taken from the series must possess the same property. Thus, for the periodic series 0 1 0 1 0 1 0 1 . . . , if we observe every other term we extract the sampling 0 0 0 0 . . . , that is, a constant series, in which the frequency of 0s is 1 and the frequency of 1s is 0. A periodic series therefore cannot be contingent according to Kolmogorov. As far as Champernowne's series goes, we can detect that it is not random by sampling it in a special way, that is, by observing only the third, eleventh, thirty-fifth, and so on, terms in the series:

$$3 = 1 + 2 \times 1\,, 11 = 3 + 2^2 \times 2\,, 35 = 11 + 2^3 \times 3\,, \ldots$$

the rule being that if the $(n - 1)^{\text{th}}$ observation is made at the nth term of the series, the nth should be made at the $(r + 2^n \times n)$th term. The sampling thus extracted from the Champernowne series is the constant series 0 0 0 . . . , from which we conclude that it isn't contingent according to Kolmogorov.

Of course the procedure adopted must not be anticipatory, which is to say it should not depend on the outcome (0 or 1) of the observation—we cannot keep or discard a draw according to whether we like it or not. In other words, knowing its position in the series but not its value, we have to be able to decide whether we'll include a term in the sampling or not. Naturally, we can't take the first *N* terms, throw out all the 1s and claim to have extracted a sampling of only 0s! In a series that is contingent by Kolmogorov's rules, each sampling that is extracted following a nonanticipatory procedure will have empirical frequencies of 0s and of 1s close to ½. These series will also successfully pass more subtle statistical tests, too technical to be related here, based on the hypothesis that they result from successive independent and equiprobable draws of 0s and 1s. In fact, from an operational point of view, they are infinite sequences of heads and tails from a balanced coin, for they will appear as such in all tests that human ingenuity can devise.

Let's look at this again. We've gone from one extreme to the other. On the one hand, we have a series engendered mechanically by a rule of iteration. Even if it seems random to an uninformed observer (this is the phenomenon of deterministic chaos), it represents a purely deterministic world, in which the future can be predicted from the past with precision. At the other extreme we have the contingent series, evidence of a world entirely devoid of meaning, in which the only rule is that there is no rule and that we can never draw any certain conclusions about the future from the past. What a surprise then to see a different form of logic arise. This world that rejects all deterministic rules fits neatly into calculations of probability. These series, carefully constructed so that we can never predict the result of any individual selection, turn out to fall within the realm of statistical predictions. The probabilistic model therefore sits at the other end of the spectrum from the deterministic model. They constitute the two poles between which our understanding of the world oscillates: as we distance ourselves from one, we draw closer to the other. A rigorously nondeterministic world must be perfectly probabilistic.

And what about the world we live in, the world in which Eyvind Kelda dies? We don't know where to situate it on this scale ranging from the random to the predictable, but paradoxically it doesn't matter. If the final reality is described by probability theory, the world will be subject to the laws of statistics. With these laws, we can accumulate independent events that are highly uncertain on the microscopic scale and obtain almost certain facts on the macroscopic scale. Determinism here is a matter of experience. Eyvind can predict with certainty that he will die in the next quarter of an hour. While it is true that the rising tide decomposes into a multitude of molecules, each of which is subject to the laws of probability, there are so many of them that the individual random destinies combine to form an inevitable event. There is no hope that the flow will stop at the foot of the rock, or, alternatively, rise up to engulf the executioner, both of which are possible, but infinitely improbable, events. This is the meaning of the second principle of thermodynamics. Even if the structured situation that we witness today is extraordinarily improbable with respect to original chaos, it exists, and from there evolution must follow according to statistical laws, that is, in an orderly manner. It is true, we do return to some kind of chaos, but not any which way. Unless we think that miracles perpetuate themselves, and that on top of a ridiculously improbable situation an even less probable scenario will arise, we are forced to predict that on the macroscopic level the most probable evolution will occur. In other words, the entropy of the system must increase. This intuition was formalized by Boltzmann in his theory of perfect gases; he was the one who showed that the growth of entropy on the macroscopic level was a consequence of what he called the "molecular chaos" hypothesis at the microscopic level.

We can't get away from determinism. Chase it out the door, by postulating total incoherence, and it comes back through the window, in the guise of statistical laws. Its nature escapes us, whether it's magical or mathematical, analogical or mechanical, but its presence seems to be a logical necessity, established irrefutably by Kolmogorov and his disciples. This is precisely Saint Anselm's famous proof of the existence of God clothed in modern garments. It went as follows: by definition, God has all perfections; the first of all perfections is to exist; therefore God exists. In scholastic language, his essence envelops his existence. Similarly, the existence of determinism, which is a question of fact, proceeds directly from its nature, which is a mathematical question.

If Saint Anselm's argument no longer convinces us today, several centuries later, it's because we're accustomed to distinguishing fact from theory. Modern man is dualistic, he makes a distinct separation between the material universe, where the former is determined, and the intellectual one, where we debate the latter. The question of a link between these two universes, which is fundamental if we want to understand our relationship to the world, barely interests him; if forced to choose, he would prefer to look at the latter as secondary to the former. This is why we are viscerally convinced that we cannot prove that something exists.

So we have good cause to look for Saint Anselm's error. We may place it in his argument's premises, that God has all perfections or that one of these is to exist. I prefer to zero in on the question of existence. In mathematics, it is well known that objects that don't exist have wonderful properties that enable us to prove anything. Thus, for example, if an irreducible fraction p/q has as its square the number 2, than its denominator q must be both even and odd. We are correct in concluding that such a fraction cannot exist, which is to say that $\sqrt{2}$ is not a rational number; this is what we call a proof by contradiction. Saint Anselm's argument is the first half of a proof by contradiction, the conclusion of which can only be the nonexistence of God. It is true that no one has yet given the second half of the demonstration that would allow us to reach the decisive contradiction; but doubts persist, and if one day they are addressed, the answer can only be negative. Let's imagine the first mathematician, a Greek or Babylonian, who studied $\sqrt{2}$ in the form of the fraction p/q. After a great many efforts, he may have proven that the denominator had to be an even number; he maintained this theory until someone else, a rival or a successor, proved that the denominator had to also be odd. Similarly, there is no guarantee that if we begin with the same premise, that God has all perfections, we couldn't logically arrive at the opposite conclusion, that God doesn't exist. At this point, the only logical conclusion would be indeed that God couldn't exist.

Many centuries have passed since Saint Anselm of Canterbury lived, and formal logic has progressed sufficiently for us to avoid these kinds of traps. Which is why Kolmogorov's analysis, which seems to establish determinism as a universal rule by the simple compelling strength of mathematical reasoning, offers us more resistance. It is certainly well-founded from a purely logical perspective. We are therefore forced to accept the Platonic idea that the infinite variety of the world is ruled for all eternity by certain theorems. Things are as they are, events follow one another, but all this is necessarily structured by mathematics. It therefore marks a limit to the contingency, for it derives from the world of theory and not of fact. We cannot imagine mathematics being any different than they are, and the physical universe itself is held in check by its laws.

We have therefore risen above the contingency of the world, and believe we have found refuge in mathematics. The physicist who discovers an unsuspected law, the engineer who calculates a structure, the economist who seeks correlations, can all testify to the power of mathematics and raise the hope that one day the world will be revealed. On that day we can say that man has mastered contingency, and that the universe is finally transparent. The physicist will have reconciled general relativity and quantum mechanics, the psychoanalyst will have found the laws of the unconscious, and all humankind will exclaim: "We finally understand. Things are the way they are because they can't be otherwise."

Then, insidious but persistent, the doubt will arise: it's too good to be true. Why accord such stature to the cerebral activities of a species trapped in a small corner of the universe? Why should mathematics escape randomness? Isn't the world to which it refers also contingent? Couldn't math have been different from what it is today on this planet?

At first glance, there is no place for contingency in mathematics. Everything is true of necessity; there is no observation of data, no authority figures who argue one way or another. Since Euclid, all mathematicians carry the same image of their science in their heads: it is based on simple axioms, combinations of which, following certain logical rules, allow us to prove other properties. Truth extends—by contagion, so to speak—from the basic axioms to everything that can be proved from them. If you can derive all of mathematics from the basic axioms, than all of mathematics will be found to be true, by a pure logical necessity, untainted with any kind of contingency, history, or chance.

It was Kurt Gödel who, in 1930, proved that this image was false. In a well-known theorem,[9] published the following year, he shows that whatever the system of axioms and rules adopted (provided there is a finite number of them), there will be some property in arithmetic that can neither be disproven nor invalidated within that system. In other words, there are mathematical propositions concerning integers that are true but that cannot be proven; they can only be noted, provided one has broad enough insights. Our demon, who at a glance can take in the infinity of integers, knows immediately if a property is true or not. But a human being does not have this faculty. He can detect an arithmetic property *de visu* only if the numbers concerned are not too large.[10] If they are too large, his only resource is to try to find the proof; if he can't find it, that doesn't necessarily mean the property in question is false. Maybe he went about it the wrong way; someone else with higher intelligence or better luck may find it.

Mathematics is full of conjectures, questions in suspension that have been waiting, sometimes for centuries, to be resolved. Gödel's theorem opens the possibility that they never will be. He goes even farther than this, allowing for the construction of a property that would be impossible to prove or disprove using the current system. This property can be taken

9. See K. Gödel, E. Nagel, U. Newman, J. Y. Girard, *Le Théorème de Gödel* (Paris: Le Seuil, 1989).

10. As an example, the largest primary number known today is $2^{21091} - 1$.

as a new axiom and added to the existing ones to constitute a new system, which in turn will bring with it an undecidable hypothesis, which can be again added to the preceding axioms to constitute a new system, and so on, indefinitely. But at each step, there is an ambiguity: if a property is undecidable (can neither be proved nor disproved) its contrary is also undecidable, so that we can choose whichever one we want to select as the axiom. Depending on whether we chose the property or its contrary, we obtain two different mathematics, each with its own perfect internal coherence, but incompatible with one another. Gödel's theorem definitively affirms the existence of an infinite number of distinct mathematics, all born of the same necessity.

Mathematics is not solely determined by logic; it contains an arbitrary element. Two positions are therefore possible. We can assert that mathematical objects, like integers, have an independent existence, so that all propositions concerning them must be true or false whether they can be proven or not. In this case, there is only one legitimate mathematics, the one that precisely accounts for all the properties of whole numbers. This is the Platonic position adopted, in fact, by Gödel himself, who in his 1931 article points out that the hypothesis he constructed is undecidable but "true," and by most logicians, who refer to a "standard model" of whole numbers, which in the final analysis means their intuition about them. We can also be pragmatic and consider that mathematical objects have only an operational existence. An undecidable question is a question without an answer, because it cannot be resolved by a physical experiment; no matter what our instincts tell us, there is no way we can state that the answer should be yes rather than no. This attitude approaches that of Niels Bohr in quantum mechanics: the question shouldn't be asked.

Most mathematicians are Platonists. This is because of the transcendent beauty of mathematical structures, which don't seem to have been created by the hand of man. It is also due to our experience as researchers, since we have more a sense of penetrating nature's secrets, of drawing eternal truths from the conglomeration of incomprehensible matter, than of crafting humble, homemade objects. Discovery brings a moment of enlightenment—we finally see things as they are, the mystery has been solved. In moments like these, we are led to reconsider the Platonic myth of the transmigration of souls, according to which souls contemplate eternal truths during their stay in the underworld before drinking forgetfulness in the river Lethe and returning to earth to begin a new cycle.

But anyone who studies the history of mathematics cannot help being overwhelmed by doubt. Naturally a few great creators have played a predominant role. What would calculus be without Newton or Leibniz, algebra without Galois, geometry without Gauss? But the history of mathematics doesn't depend solely on a few brilliant intuitions. It is part of the general development of science and technology: the growth of analysis was directed by celestial mechanics, and the book in which Gauss established the foundations of geometry was also a treatise on geodesy. Had historical circumstances been different, had there had been different needs to satisfy, wouldn't mathematics have been different? If the Earth were the only planet around the Sun and if it had no satellite, we wouldn't have spent so many centuries accumulating observations and building systems to explain the strange movements of the planets among the stars, celestial mechanics wouldn't exist, and mathematics would be unrecognizable.

We think we see a necessary itinerary, a logical development tending toward a goal. It is the illusion that the tail wags the dog, that the whole process of evolution has led us to today's observations, that the past must be read in the light of the present. Where we see steady progress along an eternal path there might only have been haphazard steps, prompted by external needs. As Antonio Machado wrote, quoted by D. Ruelle in a similar context:[11]

> Caminante, son tus huellas
> el camino y nada más;
> caminante, non hay camino,
> se hace camino al andar.[12]

We can only marvel at the singularity of our own destiny. The edifice of science, like human history, is largely arbitrary; we can only dream of what might have been. We are the survivors of a ruthless selection process, which chooses from the infinite variety of possible futures the one that will eventually occur. The events brushed aside by this faceless divinity we call chance have as much right to exist as the ones it selects, which will hereafter be part of our experience. Our greatest accomplishment is that we exist for no apparent reason and at the expense of other possibilities, which would no doubt be equally fruitful and perhaps more so. Why me? There is no answer to that question. Not surprisingly, contemplating such thoughts might provoke an identity crisis: what is this utterly contingent world, what is my life compared to so many possible lives? As the poet says:[13]

> This life
> unheard of and baroque:
> A male cell reaches out to a female cell
> and among a great number of wooers
> I happen.

11. "Are Our Mathematics Natural?" *Bulletin of the American Mathematical Society* 19 (1988): 259–68.

12. Wanderer, your footsteps are
the road, and nothing more;
wanderer, there is no road,
the road is made by walking.

(*Selected Poems by Antonio Machado,* trans. Betty Jean Craige [Baton Rouge: Louisiana State University Press, 1978].)

13. Gunnar Ekelöf, "Detta oerhörda," in *Opus Incertum* (1959).

No wonder that I doubt
that I am I

Then this society
where they all bark against each other
as in a kennel
absolutely convinced that they are they

War, human sacrifices—

I don't think I have lasted longer
than to conception:
The "spotted" birth
I, a spermatozoon, wander around wagging my tail
hunting after the egg of the world
but where is it?

Confronted with this many-faceted contingency, humanity seeks to identify the underlying determinism, that is, to give meaning to the world. As we saw, it may be found in logical necessity or in statistical regularities: the trick is to uncover it. Meaning may be an individual's personal quest or the fruit of an ancient tradition. It may also be imposed by force. Violence is pure contingency, its ultimate stage being the imposition of meaning. The tyrant doesn't merely want to be obeyed, he wants to be loved. Not only does the occupier take over the land, he also demands loyalty from the plundered populations. As Goethe puts it:

Das alte Wort, das Wort erschallt:
Gehorche willig der Gewalt.
Und bist du kühn, und hältst du Stich,
So wage Haus, und Hof, und . . . dich.[14]

14. The ancient word still makes good sense:
Succumb at once to violence!
If you are bold and don't give in,
Then risk your house and home and—skin.

Goethe's *Faust,* Part II, Act 5, trans. Walter Kaufman (New York: Doubleday, 1961).

Still, as Claude Lévi-Strauss has demonstrated, men have attributed to certain symbols the power of interpreting, even of transforming, the world. It is what we call a theory: an ensemble of fundamental elements, formal rules that allow these elements to be combined into new elements, as well as a system of correspondences between the formal universe thus created and the surrounding world. Determinism is situated in this mysterious correspondence, while the various formalisms proposed, then rejected, are as innumerable as the civilizations that created them. We have a cultural preference for the system that associates a mathematical model to an experimental verification, but other systems that are logically possible have been used. Among them is magic, along with all the occult sciences that have been passed down to us by legend, the ancient Norwegians' *seid* being just a close example, in time and in space. Documents telling us precisely what it was about are rare; the destructive campaigns of Christianizing kings played a large role in this voluntary forgetfulness.

Fortunately, we remember Egil Skalagrimson, a great Viking and great skald, a master of *seid,* perhaps the most notable personality of Nordic antiquity. The Egil Saga, itself such a masterpiece that we suspect Snorri Sturluson to have been its author, recounts how Egil used *seid* against King Eirik Bloodaxe and Queen Gunnhild. Leaving Norway after having settled his accounts with his enemies, he raised a *nidstang,* a magic stake, impaled a horse's head on it which he turned to face the interior of the country, pronouncing the following curse: "I here set up a scorn-pole, and I turn this scorn against King Eirik and Queen Gunnhild and I turn this scorn upon the landspirits which dwell in this land, so that they all fare wildering ways, and none light on or lie in his dwelling till they drive King Eirik and Gunnhild out of the land."[15]

The malediction was engraved in runes on the stake, and

15. *Egil's Saga,* trans. Gwyn Jones (Syracuse University Press: 1960).

a few years later King Eirik and Queen Gunnhild were chased from Norway and forced into exile on the Orkney Islands. They had their revenge when Egil, driven ashore by a storm, fell into their hands. They wanted to execute him immediately, but Egil had friends in the king's entourage who pleaded his case, pointing out that one doesn't kill between the setting and rising of the sun: "King, isn't it murder to kill a man during the night?" Thus Egil had a night of grace, which he used to compose a masterpiece of Nordic poetry, the first rhymed poem, an ode to his enemy's glory, which he recited the next day before King Eirik and his court. It was a recognized way for a skald in a desperate situation to save his head, and the poem has since been known by the name of *Hovudlausn*, ransom for the head.

In this story several determinisms are linked together, the practice of *seid,* the respect for custom, the poetic arts. We no longer recognize any of these as our own: neither *seid,* skaldic poetry, nor this strange form of respect that prohibits killing even your worst enemy after nightfall. But the need they expressed, to identify and if need be construct oases of regularity in the desert of contingency, is still ours today, even if we satisfy it by other means. The formalism and rigor of skaldic poetry is in no way inferior to modern mathematics, and anyone who is able to weave *kjenninger*[16] while respecting the rules of alliteration will also be able to derive theorems from one another following the rules of logic. In both cases, creativity is a bonus; it adds meaning and beauty and distinguishes the artist from the laborer.

16. The strict rules of skaldic poetry led to the replacement of many commonly used words by metaphors, the *kjenninger,* that responded better to the demands of alliteration and became more and more sophisticated over the centuries.

3

• • • • • •

Anticipation

THE MEMORY OF KING OLAF TRYGVESSON is inseparable from that of his drakkar, Ormen Lange, *The Long Serpent.* He'd had it built on the model of a drakkar that he'd brought back from Hålogaland, and that was already a superb ship. But in size and beauty, the new drakkar must have outclassed its model, which from then on was called *The Short Serpent.*

The Long Serpent's reputation had traveled abroad. During King Olaf's final expedition, his enemies, the kings of Sweden and of Denmark, as well as Earl Eirik, laid a seventy-one-boat ambush for him near the Island of Svolder, on the southern coast of the Baltic. King Olaf's fleet, unaware of the enemy presence, set sail to return to Norway. The smallest boats left first; since they were also the fastest, they were quickly out of sight. Olaf Trygvesson was left with his eleven largest ships, which a traitor, Earl Sigvalde, was to guide toward the open sea by a channel that assured a sufficient draft of water. In fact, he was leading them straight into the enemy's trap.

Svein, the king of Danes, Olaf, the king of Swedes, and Earl Eirik were at that place then, with their combined forces. It was fair weather with bright sunshine. All the chieftains now went up on the island, together with some bodies of men. They saw very many ships out at sea, and presently they saw a large and handsome ship sailing along. Then both kings said, "That is a large ship and a mighty beau-

tiful one. That is likely to be the Long Serpent." Earl Eirik answered that this was not the Long Serpent; nor was it. It belonged to Eindrithi of Grimsar.

A short time afterwards they saw another ship which was much larger than the first. Then King Svein said, "Afraid is Olaf Trygveson now, since he dares not sail with the dragon head fastened on his ship."

Then Earl Eirik replied, "This is not the king's ship. I know that ship and its sail, because it is striped. It is Erling Skjalgson's. Let it sail on. It is better for us to have a hole and gap in King Olaf's fleet than [to fight] that ship which is so well outfitted."

A while after they saw and recognized Earl Sigvaldi's ships, and they steered toward the island where they were. Then they saw three ships come sailing, one of them a large vessel. Then King Svein bade his men go on board his ships, saying that this was the Long Serpent. But Earl Eirik said, "They have many other large and stately vessels beside the Long Serpent. Let us wait still."

Then a great many exclaimed, "Earl Eirik does not want to fight now and avenge his father. It is a big shame, which will be noised abroad, that we lie here with such a large fleet and let King Olaf sail out to sea right past us."

And when they had talked about that for a while, they saw four ships come sailing, one of them a huge dragonship, all ornamented with gold. Then King Svein arose and said, "On high the Serpent is going to bear me this evening. That ship I mean to steer." Then many said that the Serpent was a marvelous, big, and handsome ship, and that it was a grand thing to have so beautiful a ship built.

Then Earl Eirik said in the hearing of several men, "Even though King Olaf did not have any bigger ships than one, King Svein would never get it away from him with the Danish fleet alone."

Then the crews hurried to the ships and removed the ship awnings. But while the chieftains were talking about this, as put down above, they saw three huge ships come sailing, and a fourth one last, and that was the Long Serpent. But as to the other two ships which had sailed past and which they thought were the Long Serpent, the first of them was the Crane, and the next, the Short Serpent. But when the saw the Long Serpent, they all recognized it, and no one contradicted that on it sailed Olaf Trygvesson. They boarded their ships and made ready for the attack.[1]

Olaf Trygvesson's enemies confront a typical decision-making problem here. They mustn't reveal themselves until the king's drakkar is in view. If they disclose their presence too soon, the king will be alerted and will have time to escape. If they set out too late, the king will have passed and they'll have missed their chance. They have to recognize *The Long Serpent.* The problem is that they've never seen it before; or rather, only one of them has seen it before, but the others don't trust him. Quickly accused of cowardice, he finds himself reduced to silence for the remainder of the debate.

The two kings have to make their decision based on *The Long Serpent*'s reputation. They have a general description of it, they know it's very large and ornate, and that it is unusual enough to be the pride of its owner, Olaf Trygvesson, whose splendid possessions were legendary. And given that naval architecture is limited in variety, and that they are very knowledgeable about drakkars, they must have a fairly clear picture of *The Long Serpent,* far more precise than what we can have today.

Nevertheless they are wrong five times. Each time they have excellent reasons for making a mistake. Each drakkar is larger and more beautiful than the previous one, and it seems each time that the summit has been reached. At each passing ship, those who believe it is *The Long Serpent,* and those who are sure it is not, face off, one side growing in confidence and the other shrinking in number. Only the appearance of the real *Long Serpent* ends the discussion. The evidence is in, no need for further discussion, they dash to their ships.

1. *Saga of Olaf Trygvesson,* chap. 101. From the *Heimskringla.*

In retrospect, the scales drop from their eyes, and they wonder how they could have been deceived by such pale imitations. It seems that the description given, a gigantic, gold-covered drakkar, could apply only to this dreamlike structure gliding in the sun. But this was not the case. Classically, those who write history create the illusion, perhaps inevitably, that its key players knew what the outcome of their acts would be. It is in fact quite difficult a posteriori to recreate the memory of a first impression. Until he had seen *The Long Serpent,* King Svein could be fooled by *The Short Serpent* and could even convince others that it was the ship they were looking for. Logical reasoning and expertise in matters of naval construction are of no help. Only the appearance of Olaf Trygvesson on *The Long Serpent* can correct his mistake.

Uncertainty is one of the basic facts of human history and of our daily lives. We are constantly forced to make decisions in contexts that we don't fully understand. Correctly evaluating a situation is crucial to judging the consequences of our actions. Our past experience can help us, as well as the body of knowledge accumulated by us and by others. And yet, we are never sure of having formulated the proper analysis. But we have to act anyway, at the risk of making decisions that, once the future has revealed the true state of affairs, may appear not only dangerous, but foolish.

We should note here that we are not necessarily talking about forays into the unknown. The situation that we are trying to analyze may be familiar. We know that what we are observing fits into a certain number of typical situations, and we are simply trying to figure out which is the right one. This enormous, elaborate drakkar belongs to King Olaf's fleet. But is it *The Long Serpent?*

We try to figure out what is going on, just as John the Baptist, from the depths of his prison, sent his disciples to ask

Jesus: "*Art thou he that should come, or do we look for another?*" (Matt. 11:2–6; Luke 7:18–23). It is part of the human condition never to obtain a straight answer to this question: "*Go and show John again those things which ye do hear and see: The blind receive their sight, and the lame walk, the lepers are cleansed, and the deaf hear, the dead are raised up, and the poor have the gospel preached to them. And blessed is he, whosoever shall not be offended in me.*" This is a recurring theme in the New Testament: you have to read the signs of the times. A reading can never be purely objective, for it involves personal and risky decisions. "*He that hath ears to hear, let him hear*" (Matt. 11:15)

The second theme is therefore related to the first: it is that faith brings revelation. For those willing to hear, the Kingdom of God is already upon us. For the community of believers, the promises are partially realized: "*And they continued steadfastly in the apostles' doctrine and fellowship, and in breaking of bread, and in prayers. And fear came upon every soul: and many wonders and signs were done by the apostles. And all that believed were together, and had all things common; And sold their possessions and goods, and parted them to all men, as every man had need*" (Acts 2:42–45). Once the belief is shared, waiting together for the Judgment induces behavior that confirms the initial choice. The prophesy comes true if enough people believe in it, as if the arrival of Olaf Trygvesson depended on the number of people who recognized his drakkar.

The capacity that certain predictions have to come true on their own, if they have enough believers, is one of the facts of life in society. If I am convinced that someone is hostile to me, I will take precautions to protect myself against his actions, and in so doing I will make him hostile even if he wasn't so to start with. If a certain political party wages a campaign on the claim that a neighboring country's treachery makes war inevitable, the future has several ways of proving it right. If this opinion gains acceptance, it may originate an arms race of uncertain outcome, or even a preventive attack, for if war is to take place, we prefer to take the initiative rather than to wait until the supposed enemy has armed himself to his satisfaction. Better still, even if the bellicose party doesn't gather a large following, the simple risk that its opinion might gain support may worry the neighboring country sufficiently for it to take military precautions, which will immediately be denounced as an implicit admission of ill intentions and lend credibility to the suspicions of hostility, thereby unleashing a deadly escalation.

It is in the economic domain that this type of situation has been best analyzed. Long ago researchers began investigating the role that agents' predictions play in the functioning of the economy, and introduced the notion of rational expectations. These are predictions that turn out to be confirmed by reality, which then confirms the agents' worldview. Thus, we may or may not believe that economic cycles are linked to solar cycles—in other words, that sunspots give rise to economic crises. But all those who believe it act accordingly, and if they are numerous or influential enough, their combined actions will lead to a crisis that will prove them right. For they will prepare for an economic crisis as the sunspot cycle nears its climax, and in so doing, by withdrawing from economic activity until the sunspot cycle is over, they will cause the crisis they expected. The theory is thereby confirmed, and the nonbelievers will have no choice but to convert to what will henceforth be a regular feature in the economy—until another opinion prevails.

The basis for economic life is uncertainty. Uncertainty with regard to the future, of course. Who knows today what an investment will produce in thirty years? The outcome depends on so many diverse factors, with some, like political change or technological progress, being outside the realm of economics; as a result our insights feel worthless and we give up on making any serious predictions. But there is also uncertainty about the present. Despite the statistics that are regularly published and discussed, no one knows what the real state of the economy is. Recessions come and go, but it is still just as difficult to recognize when one is coming or to know when to hail its passing. "Are we entering a recession?" and "Is the recession finally over?" are the two questions that alternate periodically in the news.

This is because even though they are objectively measurable, economic indicators like the inflation rate, the unemployment rate, or the trade deficit still require interpretation. They are ambiguous by themselves, just as the same symptom can apply to several illnesses. Let's put ourselves in the shoes of a manufacturer who sees a rise in prices in his sector. He must discern whether it's a monetary illusion (in which case he will simply adjust his prices to keep pace with inflation); a temporary increase in demand (in which case he must confront it with his existing means of production); or a real change in demand (in which case he must invest as quickly as possible to keep the expanded market from being absorbed by the competition). A mistaken appreciation can lead to disaster; either he will invest at the wrong time and find himself overstocked and in debt, or he won't perceive the winds of change and will be unable to maintain his place in an expanding market.

Consumers are also confronted with this type of problem: you will delay a heavy investment, a home or a car, if you expect that prices or interest rates will fall. As a result, the economic agents are constantly asking themselves what's going on. You don't need to be a scholar to know whether you're in a period of recession or expansion. What is hard to know is when a change in direction will occur, and this is the most valuable information. Correctly anticipating the end of a recession provides a decisive advantage over the competition, which may still be apprehensive and inactive.

The most remarkable point in this analysis is that the agent's evaluation is an integral part of the situation he is trying to analyze. If agents all judge that the recession will continue, investments won't be made, consumption will be delayed, and the recession will be prolonged. If on the other hand they all consider it over, liquid assets will become available for investment, a rise in demand will quickly be felt, and the long-awaited recovery will be underway. In reality of course, opinions will be mixed until one opinion gathers a majority following and the situation shifts—just as the appearance of the *Long Serpent* puts everyone in agreement.

To predict other people's opinions, to anticipate the agents' reactions before even they are aware of them, is therefore an important exercise when it comes to the economy. The worst situation is when everyone decides to wait, retreating to their cocoons until better days; the ideal situation is when the confidence that all governments aim for prevails, enabling agents, whether producers or consumers, not to see the risks they are taking.

These same mechanisms come into play on the stock market, and determine in large part the price of shares. Of course, economic considerations are key, and there is nothing like bad results to bring down prices. But on large stock markets like Wall Street, and on major stocks, the majority of purchases are sold back within the day. Clearly, a broker who holds on to his investment portfolio less than twenty-four hours is not going to be overly concerned about the value of the stocks in five years. What determines prices is the market, that is, the way thousands of brokers all over the world expect their counterparts to react.

This point has already been observed by Keynes. "Professional investment may be likened to those newspaper competitions in which the competitors have to pick out the six prettiest faces from a hundred photographs, the prize being awarded to the competitor whose choice most nearly corresponds to the average preferences of the competitors as a whole; so that each competitor has to pick, not those faces which he himself finds prettiest, but those which he thinks likeliest to catch the fancy of the other competitors, all of whom are looking at the problem from the same point of view. It is not a case of choosing those which, to the best of one's judgment, are really prettiest, nor even those which average opinion genuinely thinks prettiest. We have reached the third degree where we devote our intelligences to anticipating what average opinion expects average opinion to be. And there are some, I believe, who practice the fourth, fifth and higher degrees."[2]

We can get a taste for these subtleties by playing children's games, like "one, two, three, shoot," in which one opponent chooses odd, the other even, then both hold out either one or two fingers simultaneously for an odd or even total, or like scissors/stone/paper, in which you present either an open hand, two fingers, or a closed fist as a symbol, values being split up according to fixed rules: paper over stone, stone over scissors, and scissors over paper. The name of this game is to be one step ahead of the opponent. If he won by playing "paper" when I played "stone" the last turn, he may be tempted to play what already worked for him: this is naive, what we might call the zero degree of cunning. It supposes that I will play "stone" again, that I won't modify a losing strategy. It is more likely that I will anticipate his reaction and that I will play "scissors," thwarting his naive strategy. If he is capable of a minimum amount of reflection, my opponent will consider it likely that I will try to modify my losing strategy, and he will reproduce my reasoning. Anticipating in turn my decision, he will try to beat "scissors," which will cause him to play "stone": this is the first degree of cunning. But I can try to get one step ahead of him, thereby playing "paper," and if he thinks that I will do this, he will play "scissors." The second degree of cunning thereby returns to the zero degree; that is, the same

2. *The General Theory of Unemployment, Interest and Money*, chap. 12, V.

strategy can be adopted out of naiveté or out of subtlety. And what about degrees three, four, and more?

Think of the goalie's anguish in soccer, as he faces a penalty kick. He is not allowed to move before the ball has been hit, and his movement must be practically simultaneous with the kicker's for him to have a chance at stopping the ball. Since he won't have time to adjust, he needs to anticipate the kicker as much as possible, deciding toward which side the ball will be sent. Certain goalies try to guess according to the habits of the kicker, his level of fatigue, and the previous penalty kicks. But the opponent isn't dumb, and will try to anticipate the goalie's move. That's why other goalies don't even bother trying to outwit the kicker and let chance or intuition decide.

"'The goalkeeper is trying to figure out which corner the kicker will send the ball into,' Bloch said. 'If he knows the kicker, he knows which corner he usually goes for. But maybe the kicker is also counting on the goalie's figuring this out. So the goalie goes on figuring that just today the ball might go into the other corner. But what if the kicker follows the goalkeeper's thinking and plans to shoot into the usual corner after all? And so on, and so on'."[3]

What becomes clear in all these situations is that no method guarantees one player's success. You have to begin with the assumption that the opponents are both rational, that they have access to the same information, and that as a result each of them is in a position to reproduce the other's reasoning. If some invincible argument were to persuade the goalie to guard his right side, for example, his opponent would also have access to this argument and, if it were convincing, would anticipate the goalie's reaction and kick to the other side.

3. Peter Handke, *Die Angst des Tormanns beim Elfmeter,* trans. Michael Roloff, *The Goalie's Anxiety at the Penalty Kick* (New York: Farrar, Straus and Giroux, 1972).

This is why the best the theory of games has been able to do in the analysis of this type of situation is to introduce into the decision an element of chance. If the goalie makes his decision by flipping heads or tails, he is forever immunized against the intellectual acumen of his adversaries, and if his career is long enough he will have been right one out of two times. But he must resist the temptation of trusting his own insights, which would unfailingly trigger escalating rounds of anticipation on both sides. Even if he notices that penalties are systematically kicked toward his right side, he mustn't guard the left less often. Any change in strategy on his part would quickly be observed by the opponent, who would take advantage by placing some shots to the left. In fact, these systematic shots to one side may be intended to convince him to abandon his strategy of heads or tails, and return to a game of anticipation, with which the opponent would be more at ease.

In 1930, when the mathematician E. Borel proposed to introduce an element of chance into players' tactics, he saw it primarily as a trick which eliminated the need for cunning, a way out of the infernal cycle of escalating anticipations. In our day, we place more emphasis on the informational effects of these random strategies. The goalie leaves his opponents in a situation of permanent uncertainty: no matter how fine their reasoning, no matter how subtle their strategies, they cannot prevent him from being right one out of two times. Even inside information can't help them: if the goalie lets his girlfriend in on his plan to flip heads or tails to decide which side he'll defend, and if she gives him away, it won't hurt his chances for success. Even if he swears *urbi et orbi* that not only for this game but during his entire career he will let chance direct his moves, it will be impossible for an opponent to benefit from this information. The other side of the coin is that he will be unable to improve his success rate by taking advantage, for example, of statistical biases that he might observe in his opponents. In so doing, he would introduce a statistical bias into his own strategy (he would guard one side more frequently than the other) that an alert opponent could take advantage of. In other words, he must be able to renounce easy short-term gains to avoid giving away information that could be used against him in the long run.

We are going to reset this entire analysis in a more suitable framework, that of poker. It may seem paradoxical that we would advise a poker player to make decisions randomly, and especially that it would be the best thing to do in certain circumstances. Yet this is what we are going to do, if not for poker, at least for an ultrasimplified version of this same game. The results we obtain will remain qualitatively valid for poker itself, but the calculations are too complicated to be performed.

The game is played alone against a dealer. To begin with, you place a bet, say one dollar. The dealer must follow suit, that is to say he must wager an equal sum. He then takes a new deck of cards, breaks the seal and pulls out the sevens, which he lays out in front of himself. He shuffles the remaining cards, and invites you to cut the deck. The deck is then placed in a shoe, and the first card is yours. The dealer uses a long paddle to slide the card towards you, face down. No more cards will be drawn.

The object is to do better than the dealer's seven. The ace counts as a one; there are therefore twenty-four cards in the game below seven, and twenty-four above. There is a one in two chance that your card will be a winner, a one in two chance that it won't.

After having glanced discreetly at your card, you have two options: to pass or to up the ante. If you pass, the dealer collects the bets, and the game is over. If you up the ante, you have to bet again, always the same sum, one dollar. The dealer can then either pass or call. If he passes, you collect the bets. If he calls, he has to bet again, also the same sum. You turn over your card, and whoever has the higher card, you or the dealer with his seven, collects the bets. In either case, the game ends there.

A first analysis shows that there are exactly four strategies possible:

—up the ante with a good card and pass on a bad one,

—always up the ante,

—pass on a good card and up the ante on a bad one,

—always pass.

This last possibility can obviously be eliminated, since it inevitably means losing the initial wager. Why would the player pay for the right to participate in the game (the purpose of the initial wager) if he has decided to always pass?

The third strategy seems paradoxical. It opens up the possibility of winning if the dealer withdraws after the second wager. But if the dealer systematically calls, our loss is maximized: we lose one dollar each time we have a good card and two each time we have a bad one, which translates to an average of one and a half dollars lost per round of play.

The two other strategies produce better results. In the first we lose one dollar if the card is bad and win one if the card is good—unless the dealer ups the ante, in which case we win two dollars. If the dealer systematically passes, there is no hope of gain. As for the second strategy (always upping the ante) the result depends on the dealer's strategy:

—if he passes, we win one dollar in every case,

—if he calls, we lose two dollars on a bad card and win two on a good one, for an average of zero.

Let's put ourselves now in the shoes of an inveterate player, a casino regular who comes to play every night. Whichever strategy he adopts, in time the dealer will uncover it and adopt the corresponding defense. If the player systematically ups the ante, the dealer will call, and if the player always passes on a bad card, the dealer will never call. It therefore seems that the dealer can always reduce the player's average gain to zero.

Seen in this way, it is clear that all these strategies have the same problem: using them systematically gives them away. The dealer can use this information against the player once he has understood it, and this will allow him to limit the average gain to zero. If the player wants to increase his average gain, he will have to find a way to hide this information, a method to hide the strategy he is using.

Such a method exists: it consists in using a random strategy. If the player decides to pass on a bad card two out of three times, but up the ante one out of three times, calculations show that, on the average, he will win one game out of three—that is, his average gain will be a third of a dollar, 33 cents, which is far from negligible. What is happening, of course, is that by playing this way he deprives the dealer of the possibility of deducing usable information over the course of several games. Whatever strategy the dealer adopts, whether he systematically passes, whether he systematically calls, or whether he also adopts a mixed strategy, the average gain won't be affected, and his adversary will continue to pocket one dollar out of every three games. That is, of course, on the condition that the player remains faithful to the strategy determined at the start: he will up the ante on a bad card one out of three times. Calculations even show that this manner of playing is optimal, that is, no other strategy gets better results against all defenses.

To up the ante with a bad card has a name: it's called bluffing. A successful bluff allows you to win the game with cards that normally would not have allowed for it. The beginner is either obsessed by this possibility and bluffs all the time, or on the contrary is paralyzed by fear and doesn't dare take the risk. The experienced player realizes that it is important to be called on a bluff once in a while, so that the opponent who catches him in the act will know that he has a tendency to bluff and will be lured into following the next time he ups the ante: he wants the dealer to pass on a bad card and follow on a good one. The bluff therefore has two objectives, luring the dealer into passing on a good card and following on a bad one. To reach these goals, one has to cultivate uncertainty in the opponent, and therefore to adopt an unpredictable strategy. In order to be unpredictable, it must be random. It is not a good idea to bluff systematically, even when respecting the ideal proportion of one out of three. Deciding to always bluff on the third occasion (out of three bad cards, passing the first two times and upping the ante the third) for example, is not a good strategy. Over time, the dealer would uncover the system, and could use this information to his advantage. He would know, for example, that if he caught you in the act of bluffing (you upped the ante and he asked to see), he would be safe after that until you had passed two times, and that between now and then he could pass safely each time you upped the ante. This new strategy would diminish your hope of gain.

So we see that the dealer has an interest in continually exploring our reactions to detect cracks in our resolution, or the involuntary biases that we allow to appear in our manner of playing. For him as well, the best strategy is random. Calculations show that if the player ups the ante, the dealer should pass one out of three times and call two out of three.

For a strategy to be truly impenetrable, and remain unpredictable from game to game, it is best that it be unpredictable for the person using it, and therefore that it be random. If the player decides on a bluff as a function of the way the game is developing or according to his impression of the moment, his reasoning risks being revealed, his impression might be shared. If he throws a die, deciding that he will bet if he gets a one or a two and that he will pass if he gets three or more, he is safe from any anticipation—as long as he does this out of sight of his opponent. But he must be blindly obedient—he can't influence the outcome or misread the die, like Rabelais' Judge Bridlegoose, who passed sentences by rolling dice.

"What I do is exactly what you do, gentlemen, and as judges in general do, to whom our laws require us to defer: ut no. extra. de consuet. c. ex literis, et ibi Innoc. *First I make sure I have carefully seen, reseen, gleaned, regleaned, combed through and skimmed the complaints, motions for adjournment, summonses, charges, inquiries, pretrial pleadings, factual documentation, legal citations, allegations, defenses, demands, counterdemands, replies, duplicates, triplicates, records, challenges, legal arguments, exceptions, redeterminations, evidentiary statements and restatements, accusations, motions to change venue, official letters, demands for official copies, motions to disqualify, orders to show cause, jurisdictional challenges, allegations of conflict of interest, motions to transfer to another court, statement of conclusions, summaries of motions pending, requests for appointments, appeals, admissions, sheriffs' reports, and all the other sweets and spices emanating from one side or the other, as to be sure the good judge must always do: see* no. Spec. de ordinario sec. iii et tit. de offi. omn. ju. sec. fi. et de rescriptis praesenta, sec. i.

"Then I push all the defendant's sacks of paper off to one end of my desk and assign him the first throw of my dice, just as you do, gentlemen: as in not. 1. Favorabiliores ff. de reg. jur. et in c. cum sunt eod. tit. lib. vi, *which says:* Cum sunt partium jura obscura, reo favendum est potius quam actori, *When it's not clear what the rights of the parties are, it is necessary to favor the defendant over the plaintiff. Having done this, I also push aside the plaintiff's sacks of paper, just as you gentlemen do, though of course to the opposite end of my desk,* visum visu, *facing each other. Because* opposita juxta se posita magis elucescunt, *When opposing things are set against one another, they become easier to understand,* ut not. in 1. i sec. videamus, ff. de his qui sunt sui vel alie. jur. et in 1. munerum. i. mixta ff. de muner. et honor. *In the same way, and at the same time, I throw the dice for him, too."*

"But my friend," asked Trinquamelle, "how do you deal with the vague claims put forward by the parties to the action?"

"Just as you do, gentlemen," answered Bridlegoose, "that is, by

determining which of the parties has brought in a large sack of papers. And then I rely on my small dice, as you do, gentlemen, in observance of the law Semper in stipulationibus, ff. de reg. jur. *and that law written out in versified capital letters,* g. eod. tit.:

Semper in oscuris quod minimum est sequimur,

In uncertain matters, always strive for minimal consequences, as accepted into canon law in c. in obscuris eod. tit. lib. vi.

"And I have some other nice fat dice, which I use, as you gentlemen do, when the case is clearer—that is, when there's less in the sacks."

"And when you've done that, my friend," asked Trinquamelle, "how do you make your decision?"

"Just as you do, gentlemen," answered Bridlegoose. "I give judgment in favor of whichever party is first awarded it by a throw of the judicial, tribunal, praetorial dice. That is what our laws require: ff. qui p. in pig. 1. potior. leg. creditor. C. de consul., 1. i. Et de reg. jur. in vi. Qui prior est tempore potior est jure, *The law is favorable to the person who gets there first."*[4]

We are not making fun of Bridlegoose, whom Rabelais actually treats with great indulgence. Because in reality, over the centuries, judges and priests, generals and dictators, kings and emperors have decided matters of life and death on a roll of the dice, the flight of a bird, the fall of a horse, the appetite of a sacred chicken, the intestines of sacrificed animals, the birth of a monster, the passing of a comet, the grounds in a coffee cup, the mutterings of the Pythia, the embers in a hearth, a cloud of smoke, the lines of a palm, a sneeze, a cry, a dream. Isn't humanity's frenzy to consult fate and interpret signs an expression of our desire to penetrate the intentions of a higher authority? Are we not engaged in a game against a

4. Rabelais, *Gargantua and Pantagruel, The Third Book,* chap. 39, trans. Burton Raffel (Norton: New York, London, 1990).

Player whose greatest feat is in dissimulating not only his strategy but his existence and what he expects of us?

Or are we simply afraid of risk, and are we seeking desperately to unburden ourselves of the responsibility which weighs on our shoulders? Certainly for an ordinary decision, when one still has an opportunity to change one's mind, we can afford to make a mistake, and we can appreciate the statistical successes of random strategies. Today's loses will be compensated by tomorrow's gains. But who can know how alone the Athenians felt before the battle of Salamis? The other Greek cities surrendered to the Medes with hardly a fight. The Athenians alone abandoned their city, setting off to sea despite their inexperience; they took their wives and children to the island of Salamis, and now they are in the strait about to confront an army and fleet ten times larger than their own. If their decision was wrong, there won't be a second chance; the men will fall to the swords, the women and children will be taken as slaves, and history will forget the Athenians and their city.

Even if such extreme examples are rare, we are constantly making decisions, sometimes major decisions, in circumstances that come only once. A marriage or the choice of a career lasts a lifetime; an investment or the purchase of a home involves large financial risk. Each one of these decisions is unique of its kind and full of uncertainty: we know when we make them there will be no going back, yet we don't have all the elements of the choice at our disposal. It is natural for us to be tempted to assemble all the information possible, even if it has only a distant relationship to the problem, like the appetite of sacred chickens or the appearance of sunspots. And we can always benefit from self-fulfilling prophecies. If the couple is convinced that their marriage began under the right star, their conjugal life may benefit. If all the investors are persuaded that sunspots provoke economic crises, they will take appropriate measures as soon as the spots become abundant, provoking the crises they predicted. The oracles and all forms of sorcery play an important social role by orienting individual expectations towards an outcome favorable to the collectivity, and by thus augmenting the chances that these favorable outcomes will occur. When the Chaldeans besiege Jerusalem, the prophets run around the town predicting the arrival of the Egyptians and the lifting of the siege, thereby playing their social role. You have to be Jeremiah to dare preach defeatism: "He that remaineth in this city shall die by the sword, by the famine, and by the pestilence: but he that goeth forth to the Chaldeans shall live; for he shall have his life for a prey, and shall live" (Jer. 38:2). We can understand the violent reactions of the commanders and of King Zedekiah, who threw Jeremiah into a cistern. But it is precisely his nonconformity that is the best sign of his being chosen by Yahweh.

This is in fact the ultimate meaning of Bridlegoose's fable, since it concludes with the story of the Fifth Lateran Council (chap. 41). When the dispute is exhausted after a long, prosperous life, and the parties have gone broke in trials and appeals, when they've exhausted their friends' and families' credit, and anger has long since given way to lassitude, the moment of reconciliation arrives. The parties now want peace. The only thing that holds them back is the indignity of having to take the first step. At this point an external sign—judgment being passed, the mediator's intervention—is enough to win over the two adversaries and for the entire dispute to be buried. The final decision becomes unimportant, as long as it's a decision. Bridlegoose can roll his die, big or small, in his chamber before the stacks of texts. The trial has reached perfection, the initial grievances have long been forgotten. Bridlegoose's oracle creates the formal conditions that allow the parties to be reconciled, and the tear in the social fabric to be repaired. To let chance decide would be absurd if it were a matter of applying logical or moral criteria. It becomes perfectly reasonable when it is a matter of giving a social signal, like two motorists who arrive simultaneously at an intersection and accept that the color of the lights decides on the order of passage.

In a famous lecture on the tales in *A Thousand and One Nights,* Borges tells a story that I was unable to find in my edition. It's the story of a resident of Cairo, who in a dream receives a call to go to Ispahan, to a certain mosque where a treasure awaits him. The dream recurs several times, so that the man undertakes the voyage. It is no small affair; he travels from caravan to caravan, at the mercy of every kind of robber. When he finally arrives in Ispahan he is exhausted and penniless. He goes to spend the night at the mosque in question. It turns out to be a thieves' den, which the police raid that very night. Abused and beaten, the man from Cairo is led before the kadi and called upon to explain his presence. He tells his story, whereupon the magistrate bursts into side-splitting, uncontrollable laughter. As soon as he gets hold of himself he wipes his eyes and tells him the following: "Naive and credulous stranger, three times I have dreamed that I must go to Cairo, to a certain street, that I will find a house there, and in this house a garden, and in this garden a basin, a sundial and an old fig tree, and under the fig tree a treasure. I never paid any attention to it, and I see today that I was wise. Here is some money, take it, go home, and beware of believing in dreams sent by the devil." The man from Cairo thanks him, goes home, walks into his garden, digs under the fig tree between the basin and the sundial, and finds the treasure.

The beauty of this story is that both the kadi and the traveler can congratulate themselves on the wisdom of their own judgment. Their analyses, which are diametrically opposed, are fully confirmed by the events. Until his dying day, the kadi in Ispahan will laugh at those naive enough to undertake such a long voyage in search of a nonexistent treasure. And the man from Cairo will rejoice his entire life at having believed his dream. Each in his own way had perfectly anticipated the events.

4

. . . . . .

Chaos

inar Thambarskelfir was on The Long Serpent *in the compartment forward of the stern. He shot with bow and arrow. He was the best shot anywhere. He shot at Earl Eirik, and the arrow hit the top piece of the rudder above the Earl's head and sank in all the way up to the socket. The Earl looked at it and asked if they knew who was the archer; but straightway there came another arrow, and so near to the Earl that it passed between his side and his arm and went into the headboard behind him so that the point came out on the other side of it. Then the Earl said to the man some call Finn (some say he was Finnish)—he was a great archer—"Shoot that big man in the forward compartment!" He shot, and the arrow struck the bow of Einar in the middle, at the moment when Einar drew his bow for the third time, and the bow burst in two.*

Then said King Olaf, "What cracked there with such a loud report?"

Einar answered, "Norway, out of your hands, sir king."

"Hardly so great a break," said the king. "Take my bow and shoot with it"—and he flung his bow over to him.

Einar took the bow and at once drew the head of the arrow behind it and said, "Too soft, too soft is the king's bow," and threw the bow behind him, took up his shield, and fought with his sword.[1]

THE BATTLE OF SVOLDER ended in defeat. After a difficult struggle, the defenders of *The Long Serpent* were outnumbered. Those who didn't die fighting jumped overboard to avoid falling into enemy hands. Among them was King Olaf Trygvesson, whose body was never found and whose disappearance remains a mystery. Einar Thambarskelfir survived the massacre. He was only eighteen years old at the time, and if the saga bothers to mention this, it's because it was a unique honor to have been chosen so young to be part of the king's personal guard. A colorful character, he would become one of Saint Olaf's most faithful companions, and would die only years later, assassinated with his son in an ambush set by King Harald the Harsh.

Norway is divided between the three victorious allies, who leave the scene of battle triumphantly aboard Olaf Trygvesson's magnificent vessels, Earl Eirik at the helm of *The Long Serpent*.

The outcome could easily have been otherwise. Einar Thambarskelfir's arrow passed a few centimeters from the head, then from the chest, of Earl Eirik. Judging from the way it was launched, it would have caused pretty serious damage had it hit him. On the other hand, Finn's arrow hit Einar's bow just as he was taking aim for the third time, a miracle of precision in space and time. What luck for one, what misfortune for the other. If Earl Eirik had been killed, *The Long Serpent* would have escaped, and the other side would have been victorious. Olaf Trygvesson would have returned to his kingdom, and Norway would have avoided a long period of trouble that was to conclude with the coming to power of Olaf Haraldsson. This personage would no doubt have remained

1. *Saga of Oláf Trygvesson,* chap. 108. From the *Heimskringla.*

obscure, he would not have become Saint Olaf, and we wouldn't have the masterpiece of Saint Olaf's saga today.

We marvel at the fact that a few centimeters' deviation in the course of an arrow could change human destinies and decide the fate of a kingdom—in the final analysis, this translates into a few tenths of a millimeter to the left or to the right in the position of the fingers on the bow, and a few tenths of a second earlier or later in releasing the arrow. A good marksman is one with the sensitivity needed to make such adjustments. It is because an infinitesimal displacement of the hand incurs a significant modification in the course of the arrow that archery is possible and interesting. It requires precise gestures and finely honed perceptions, which a novice cannot achieve but which can be acquired over time by methodical and assiduous training. It is like a second nature acquired through rigorous practice, which opens new perceptions. Such discipline is hardly required to throw a stone; but stone throwers are less exact than archers, unless they use slings, a weapon that also requires training.

It was Pascal who remarked that matters of great consequence are decided by intangible factors: "Cleopatra's nose: had it been shorter, the whole aspect of the world would have been altered."[2] Indeed, if Anthony's fleet fell into disarray at the battle of Actium, just when victory was within reach, it was because the flagship could be seen fleeing the battlefield in pursuit of Cleopatra's galley ship, which was abandoning such arduous combat. Would a Roman world governed by Anthony have been measurably different from the one governed by Augustus? We may have some doubt about this, but we should also consider that the intellectual flowering that marked the century of Augustus was closely linked to the personalities of Augustus and of his friend Maecenas, and that without this

2. Blaise Pascal, *Pensées: Thoughts on Religion and Other Subjects,* fragment 162 (New York: Washington Square Press, 1965).

accident we would have neither Virgil nor Horace, nor so many other creators who have left a profound mark on our civilization today.

The lamented Isaac Asimov wrote an excellent short story, "Spell My Name with an S," which first appeared in 1958. The hero, one Marshall Zebatinsky, is a second-rate physicist languishing in a third-rate U.S. army laboratory whom some kind of guru promises a change for the better in his fate if he consents to changing one letter in his name—the Z should become an S. Marshall Zebatinsky heeds this advice and goes to an enormous lot of trouble to become Marshall Sebatinsky. A few months later he is offered a chair at a famous university, an offer beyond his wildest dreams.

The hidden side of the story is that his request for a name change, by its very modesty, had attracted the attention of the FBI. They would have understood that the burden of such a name should cause him to adopt another one, or to make some drastic modification. But to change a single letter, to go from one unpronounceable name to another, was strange enough to attract the attention of the authorities. The file makes it to counterespionage, which decides to research homonyms in the East, and which discovers in its files a Soviet specialist in a slightly neglected domain of nuclear physics. They undertake a systematic search, which brings to light the fact that all the known specialists in this specific domain have disappeared from circulation in the course of the year, no doubt recruited by a clandestine laboratory. Working backward, little by little, they bring to light a vast Soviet military effort and avert a third world war.

The climax of the story is its conclusion. Of course, the guru is actually an extraterrestrial. He just won a bet with one of his friends; he had to make something of primary importance happen (the destruction of the planet has been averted) by a minor action (changing a letter in the name of an obscure person). The loser sadly admits defeat. He proposes double or nothing: reverse what you have done, and obtain the destruction of the planet by some minor act. The challenge is accepted, and it is the end of the story.

What this story shows, and what is underlined by Pascal's meditation, are the major consequences that slight modifications can produce in the normal unraveling of a temporal process. The phenomenon is well known in mathematics under the name of *exponential instability.* We know, for example, that in meteorology the magnitude of a disturbance doubles every three days if nothing interferes with its development. In mathematical language, the equations that regulate atmospheric circulation, on which weather depends, have the property of exponential instability. Any initial condition, any atmospheric state on the surface of the globe (pressure, temperature, humidity), corresponds to a predetermined future evolution, resulting from a calculation that may be out of our reach to perform, but which is not based on chance. If we modify this initial condition very slightly, if for example a butterfly bats its wing or someone lights a candle, this infinitesimal change will be of little consequence in the moments or days that follow, meaning that we will hardly be able to distinguish the difference between the initial state of the atmosphere and the modified state. But the effect may amplify over time at an exponential rate; if it doubles every three days, it will be multiplied by 1,000 every month, 1,000,000,000 every two months, and 10^{36} every year. This is an enormous number, which means that the flapping of a butterfly's wing or the flame of a candle can cause a cyclone at the end of a year, in the sense that in a test atmosphere, in which everything else remains the same, if this butterfly or this candle hadn't existed there wouldn't have been a cyclone at that moment.

This doesn't mean, of course, that we have to be afraid of butterflies, or that candles are bad for the environment. Usually, the disturbance is balanced out by others. It is likely that the slight breeze created by a butterfly or by a candle will be diluted among the myriad others that affect the environment at every instant. But disturbances can sometimes congregate, and if conditions are favorable the slightest breeze will be enough to send the atmosphere into a complex evolution, the long-term result of which will be a cyclone or some other major catastrophe, just as a loose rock will yield to a slight shove and cause an avalanche. This means that if we want to predict what's going to happen, if we want to know what the weather will be like one year from today, we have to take everything into account, from the butterflies flying in the Amazon jungles to the candles burning in churches. As a result it becomes impossible to calculate in practice: how can we gather and process such a quantity of observations? This is why, even though meteorological science has some of the most powerful computers available at its disposal, it is unable to predict the long-term weather.

All these ideas are currently very fashionable, but they can be traced, like many others, to John von Neumann. The chief scientific adviser to the American government during the war, he thought a great deal about the potential of the electronic calculator (the first models of these calculators were built at Los Alamos under his direction). One of the main strategic problems was in correctly predicting the weather—we know the critical role this prediction played in the launching of Operation Overlord—and the new capabilities of automatic calculators were quickly in demand. Von Neumann soon realized the limits of this approach, and understood that the sensitivity of equations with regard to the initial conditions would always prohibit any precise long-term prediction. But since he had a great mind, he also drew another, far more original, conclusion: it was that this very instability might allow us to direct the weather. After all, if the beating of a wing or a candle's flame can have such tremendous repercussions, why not provoke them sooner? Perhaps it would be easier to control the weather than to predict it.

When we drive a car, we take advantage of a similar instability. To understand this, all you have to do is let go of the steering wheel: the car curves off course, slightly at first, then more and more sharply, until it leaves the road or goes into a skid. That is why one finger is enough to drive a car that weighs a ton or more. Unfortunately, we have no mechanism that would allow us to control the atmosphere as directly. The effect of a small disturbance is felt only long afterward, a year later if we descend to the level of the flapping of wings. If we want to obtain more observable, and thus more easily controllable, effects, we need much greater disturbances to the atmosphere, on the order of thermonuclear explosions.

In conclusion we should note that all these phenomena depend on the time scale. On the scale of one year, exponential instability makes it impossible to predict the weather. No calculation is precise enough to tell us whether it will rain in Paris one year from today. If we look at things on a more limited scale the problem disappears: we can predict tomorrow's weather with some success, and you don't need to be a weatherman to guess the weather in a minute or even in an hour. On the other hand, if we look at things on an even larger scale, that of a millennium, weather becomes climate, and exponential instability appears in a different light. Indeed, on this scale, it is no longer a matter of predicting the weather but of discerning certain regularities which geographers classify and study under the name of climate. Oceanic or continental climate, temperate or tropical climate, equatorial or polar climate, the list is familiar; we never even question the existence of these major regularities, the return of the same conditions during the same periods of the year. When the weather seems to deviate, when the years tend to be warmer and dryer than usual, we consider that the climate is off track, and that the causes must lie in some external agent, like atmospheric pollution or an atomic bomb.

And yet why should we expect regularity on the scale of several millennia when we don't find it on the scale of the day or the year? In the course of a single season, some days are sunny while others are cloudy, some are dry, some are rainy. Is it unreasonable to consider that it might be the same on all scales, that warm spells will naturally follow cold spells, without requiring the intervention of an external agent? Did solar cycles cause the ice age, or did it simply result from the internal evolution of an exponentially unstable system? Does the very notion of climate have any meaning beyond a few centuries? Why do we insist on seeing regularity where there manifestly isn't any? Why do we always try to explain deviations relative to a norm that exists only in our minds?

Let us look at this planet: from the astronaut's point of view, a blue jewel floating in an ebony sky. On its surface we see the play of white clouds, bearing rain and attesting to atmospheric currents. This earth is unique in more than one way, notably because we can do no more than follow the never-ending dance of the clouds, without being able to predict their formation very far in advance. Geologically, the ocean levels vary and the glaciers move in a random manner. And yet, how paradoxical that this almost complete unpredictability is accompanied by great structural stability. Year after year, even if it sometimes comes late, the monsoon returns to bathe the shores of Asia, from India to China. The anticyclone of the Azores moves according to the seasons or the years, but doesn't disappear: at any given moment we can be sure of finding a great anticyclone west of Gibraltar.

If the Creator were to place another Earth before our eyes, constructed according to the same rules as for this one, we would see a spectacle in every way similar to the one we observe here. True, the weather would be no more predictable than our own, and the succession of climates would be no less erratic, but at each instant the sight we beheld would be familiar to us. A beautiful day on that planet would resemble a beautiful day on earth. We would find the succession of seasons with its train of meteors, the monsoon in Asia, and anticyclones in the Azores, provided of course that the distribution of the continents was about the same. To put it another way, similar geographical conditions would produce similar climatic effects, whereas it is not true that similar atmospheric conditions at any given instant will produce similar atmospheric conditions one year later. Exponential instability prohibits any long-term quantitative predictions, but it does not exclude qualitative predictions, even in the distant future.

The reader may wonder why I'm making so much of this. It seems pretty obvious that the atmosphere is an extremely complicated system, in which areas separated by great distances or altitudes wind up affecting one another. Added to this is the influence of the terrestrial surface and of the interplanetary void, with which the atmosphere is in constant contact. No wonder that such a complicated system would have such complex behavior. The uncertainty of any prediction derives simply from the impossibility of mastering all significant parameters.

But this simply isn't true. It's not because the system is complicated that its behavior is unpredictable. There exist very simple systems whose behavior is just as unpredictable. In fact, we owe it to the meteorologist E. Lorenz to have reduced the many equations governing the evolution of the atmosphere to three, and to have shown that the reduced model preserved the near infinite complexity of the original.

Exponential instability and the difficulty it implies in making predictions are common phenomena which manifest themselves in a great variety of situations, from simple to complex. To fully understand it, it is best to study a simple example. We will abandon meteorology for the moment, with its thousands of variables linked by differential equations, to focus on systems described by a single variable, X. One number will thus suffice to entirely define the state of the system: it is the value adopted by the state-variable at the moment under consideration.

To continue our attempt at simplification, we will consider that time is discrete, that is, the variable time can be represented only by integer values $n = 1, 2, 3 \ldots$, the value $n = 0$ indicating the initial instant and the negative values $n = -1$, $-2, -3, \ldots$ representing moments in a more or less distant past. The evolution of the system over the course of time is therefore completely described by the series of values X_n that the state variable X assumes at every instant of n in the past (for $n < 0$) and in the future (for $n > 0$), a series that is therefore doubly infinite since the index n takes on every integer, positive and negative. A system is deemed deterministic if the state X_n at moment n is related to the preceding state X_{n-1} by a relationship of the type:

$$X_n = f(x_{n-1}).$$

The function f is the law of the system. Its very presence guarantees that the entire history of the system and its entire future are inscribed in the initial state X_0. Indeed, a simple application of the preceding formula gives us successively $X_1 = f(x_0)$, then:

$$X_2 = f(X_1) = f(f(X_0)) = f^2(X_0),$$

and so on, following the general formula $X_n = f^n(X_0)$. We say that Xn is the nth *iteration* of X_0.

As simplistic as they may seem, these models with a single state-variable and with discrete time nevertheless produce phenomena that one would have thought characteristic of models with several state-variables and with continuous time. Exponential instability is one of them. Let's compare, for example, the two systems:

$$X_n = X_{n-1} + 10,$$

and

$$X_n = 10 \times X_{n-1}$$

They lead to the two explicit formulas, giving the state at time n as a function of the initial state:

$$X_n = X_0 + n \times 10,$$

and:

$$X_n = X_0 \times 10^n,$$

which manifest two very different behaviors for two systems that are perfectly deterministic.

In the first case, we pass from one state to the next, adding a fixed quantity to the descriptive variable. Such a system indefinitely perpetuates the precision of any observations. If for example an error of 0.001 has been made on the initial state, this same error will taint the successive states for as far as one cares to travel into the future or as far back as one reaches into the past. The explicit formula $X_n = X_0 + 10^n$ shows that the error in X_0 is perpetuated on X_0 with neither reduction nor amplification. On the other hand, in the second case, the explicit formula $X_n = 10^n X_0$ shows that the errors are multiplied by 10 upon each iteration, and are therefore rapidly amplified as one travels through time. An error of 0.001 on the initial value X_0 becomes an error of 1 by the third iteration, 1,000 by the sixth, 10^{n-3} by the nth.

One way to visualize this is to imagine that the state-variable X represents the position of a point on a circle, measured by the number of rotations from a point of reference A. Thus, the value of the state-variable $X = 0$ signifies that the representative point M is at A. The value $X = 1/4$ signifies that M is derived from A by making a quarter turn forward (counterclockwise), $X = 1/2$ indicates the point diametrically opposed to A on the circle, and $X = 1$ gives us point A again. It is crucial to note that two values of X that differ by 1 correspond to positions that differ by a complete rotation, that is, they in fact represent the same point on the circle.

The two deterministic laws that we have presented therefore behave very differently. In the first, $X_n = X_{n-1} + 10$, the representative point M remains fixed: ten complete revolutions occur at each iteration, that is, it always ends up in the same place. This law is therefore merely a complicated way of describing immobility: the system's representative point remains where it was at the start. Of course, a mistake in marking the point, or a small displacement of the initial point, will remain the same at each instant.

This is not true of the second system. The law adopted now translates into a real displacement from the representative point, a displacement that depends both on the initial state and on the instant at which it is being considered. A point situated at A at the beginning ($X_0 = 0$) will remain there indefinitely ($X_n = 0$). If on the contrary we begin at point B, which is diametrically opposed to A on the circle ($X_0 = 1/2$), we will find ourselves at A after the first iteration, and we will remain there ($X_n = 10^n/2$ is an integer once $n \geq 1$). Generally speaking, if the initial value X_0 has a finite decimal expansion, that is, if it is written $X_0 = a_0.a_1a_2a_3 \ldots a_N000 \ldots$, all the digits after a_N being zeros, then the representative point M will remain indefinitely fixed at A after the nth iteration.

On the other hand, if the initial value X_0 requires an infinite decimal expansion, the movement will be much more complicated. Specifically, if the initial position is given by $X_0 = a_0.a_1a_2a_3 \ldots$, with an infinite number of digits after the period, the position at instant n will be obtained by moving the period n places to the right. Thus, at the first iteration, we will obtain $x_1 = a_0a_1.a_2a_3a_4 \ldots$ Of course we can ignore the two digits before the period, which correspond to a whole number of turns and therefore don't influence the position, which is determined by the remaining digits: $0.a_2a_3a_4 \ldots$ Similarly, position at instant n will be given by $x_n = 0.a_{n+1}a_{n+2}\ldots$; the most important indication is given by a_n+_1, which determines the position of the circle within 1/10. From this point of view, our system functions like a magnifying glass, or rather like a microscope, which at different levels of magnification reveals an increasing amount of detail.

Each additional observation reveals one more decimal digit. The initial position is essentially given by a_1, the first digit after the period; but this number becomes irrelevant as soon as we have the next one, a_2, which then determines the position. At each stage, we have to look at digits further and further from the period to get an idea of the position of the representative point: to know the system state at instant n, we must know the initial state x_0 until $(n + 1)$ digits after the period.

The immediate consequence is that if we want to be able to predict the entire evolution of the system in the future, we must know all the digits after the period. This is manifestly unrealistic. The mathematician can write that he places a point at the distance 1/2, or $\sqrt{2}$, or π; it's a purely intellectual operation, a geometric abstraction in which we manipulate intangible points, with no shape and no depth. But, in reality, the precision of observations cannot be unlimited. A mathematical model like the one we proposed therefore no longer has any physical significance below a certain scale or beyond a certain duration; the long-term behavior is thus determined by fluctuations situated outside of the model considered.

Worse still, even when we can continue to follow the system, we observe a regular loss of precision over time. If we know the state with N decimals at the initial instant $t = 0$, we won't know more than $N - 1$ at the following instant $t = 1$, and $N - n$ at instant $t = n$, until we lose all information at instant $t = N$. In other words, the errors are multiplied by 10 at each step, until all the information that we initially began with is affected. This is exponential instability. In a system that has this property, whether it is one-dimensional and simple, like the one that we have just described, or multidimensional and complex, like the atmosphere, it functions as a disclosing device, gradually revealing the information contained in the initial condition, which was inaccessible upon direct observation because of the necessarily limited precision of our means of measurement. In the elementary case we've discussed, the nth observation reveals the $(n + 1)$th decimal of the initial position. In the case of meteorology, the weather we observe in a year will reveal information about the state of the atmosphere today, information that is situated on a scale too minute for us to observe it directly.

We can also take another point of view, and consider that all information is necessarily finite, and that it would be a waste of time, for example, to try to go beyond twelve significant digits, which is the current limit to the precision of physical constants. At that point, we are no longer revealing but creating information. If we can know only the first twelve decimals of the initial state, $X_0 = 0. a_1 a_2 a_3 \ldots a_{12}$, the twelfth observation $x_{12} = 0.a_{13} \ldots a_{24}$ brings truly new information. Similarly, it would be very difficult, if not impossible, and decidedly uninteresting, to draw from meteorological conditions that will occur a year from now the information that we are lacking about the state of the atmosphere today. It is better to consider that there is a creation of information over time.

According to the point of view we adopt, there is revelation or creation of information. In either case, a piece of information essential for the future evolution of the system is not available. We may consider that this deficiency is due to our own inadequacy, in which case there is a latent revelation of information, or on the contrary that it is in the nature of things, which means that the twelfth decimal that appears at each observation is a creation *ex nihilo*. We can express this conveniently by saying that the future evolution of the system depends on the state that we observe today, and on chance.

In this way, we blow a cloud of chance over information that we have given up on discovering. In the first interpretation, it is a matter of relative, human chance, the way a player cannot sufficiently control his dice to make them land on the numbers he'd like. In the second case we are talking about essential randomness, a natural chance, the kind that we recognize in quantum mechanics. But whether there is a creation or a simple revelation of information, the same question arises: at what rate? That's what the entropy of the system measures.

The concept of entropy has had plenty of ups and downs. We are referring here to the ideas of Shannon and Kolmogorov, that is, we are looking at it from the point of view of the theory of information and of dynamical systems, and are not attempting any interpretation in terms of order and disorder. For us the entropy of a dynamical system is a number that measures the speed at which it creates (or reveals) information.

For circular one-dimensional systems, like the two just mentioned, knowing an additional decimal allows us to localize the point in question with ten times more precision, in the sense that among all the points whose decimal development begins with $0.a_1a_2 \ldots$ a*N*, only one in ten will have as its ($N + 1$)th decimal a given digit, 3, for example. In other words, numbers whose decimal development begin with $0.a_1a_2 \ldots a_N$, occupy a small interval, divided into ten equal subintervals corresponding to the ten possibilities for a_{N+1}; to say that the ($N + 1$)th decimal is a 3 eliminates nine of these ten subintervals. Thus, revealing successive decimals divides the range of possibilities each time by 10. We agree that this gain by a factor of 10 corresponds to an entropy of 1. Thus, revealing two additional decimals corresponds to a gain by a factor of $100 = 10^2$, or an entropy of 2. On the other hand, the status quo, that is, gain by a factor of $1 = 10^0$, corresponds to an entropy of 0.

The entropy of the system measures the average precision afforded by each new observation. Thus, for our first system, $x_n = x_{n-1} + 10$, which we have already observed corresponds to immobility, pure and simple, entropy is 0, which means that a new observation brings no additional information. But for the second, $x_n = 10 \times x_{n-1}$, each new observation provides an additional decimal, which means that the entropy of this system equals 1.

For multidimensional systems, the situation is complicated by the fact that they may simultaneously exhibit characteristics of stability and of exponential instability. Let us see why.

We should first look at how a one-dimensional system described by the law $x_n = x_{n-1}/10$ would appear on the circle. At each step, the state-variable (the number of turns starting from A) is divided by 10, which means that it rapidly tends toward zero, and that we always end up at point A, corresponding to the state $x = 0$. This point is what is called an attractor. Whichever departure point M_0 is chosen, the representative point M_n at instant n tends ineluctably toward A as time passes. This is a phenomenon of exponential stability (no longer of instability), which corresponds to a loss (no longer a gain) of information: no matter where we start, we will always end up at point A. Any information as to the initial position turns out in the long run to be redundant.

We can easily show this same phenomenon in a multidimensional system. Let's imagine, for example, working on a (two-dimensional) disk instead of a (one-dimensional) circle. In other words, each point M on the disk corresponds to a state of the system: we are therefore dealing with a two-dimensional system. As in the case of a one-dimensional system, we next specify the law defining the succession of states. Here, we agree that if the state at instant n is at point M_n, the state at instant $(n + 1)$ will be at the point M_{n+1}, in the middle of the segment OM_n, where O is the center of the disk. Formally, the system's law will be written $OM_{n+1} = OM_n/2$, which is the two-dimensional counterpart to the laws that we have written until now. The behavior of the system is easy to describe: whatever the point of departure M_0, the representative point M_n indefinitely approaches the center O as time passes, that is, as n increases. We express this fact by saying that the state represented by the point O is an attractor for the system.

It is easy to construct an attractor for a three-dimensional system in a similar way. You take what geometers call a torus, a solid in the form of a tire, a full circular ring, and compress it around its core. More precisely, the torus has a symmetrical axis (the hub of the wheel), and the cross sections that pass through this axis are disks. The centers of these disks span a circle, which is symmetrical with respect to the axis: this is what we call the core of the torus. In contracting each of these disks around its center, as illustrated, we define a determinist law for the entire torus, which finds itself compressed around its core. This core then turns out to be an attractor for the resulting three-dimensional system: the representative point M_n converges toward the projection of the initial point M_O onto the core of the torus. The attractor is a bit more complex than in the preceding case, since it's a curve instead of a point, and this complexity carries over into the movement, since the final state which the system tends toward depends on the initial state. Two different points of departure correspond to two different projections on the core of the torus.

Now we have seen examples of unstable systems and of stable systems, each one possessing very different properties. One of the most fascinating aspects of multidimensional systems is that they can simultaneously present both of these properties—even though they seem mutually exclusive. Indeed, since several dimensions are possible, we can imagine determinist laws that would expand in certain directions and contract in others. This leads to spectacular geometric constructions, which Stephen Smale was the first to present. He is

A Simple Attractor in the Torus

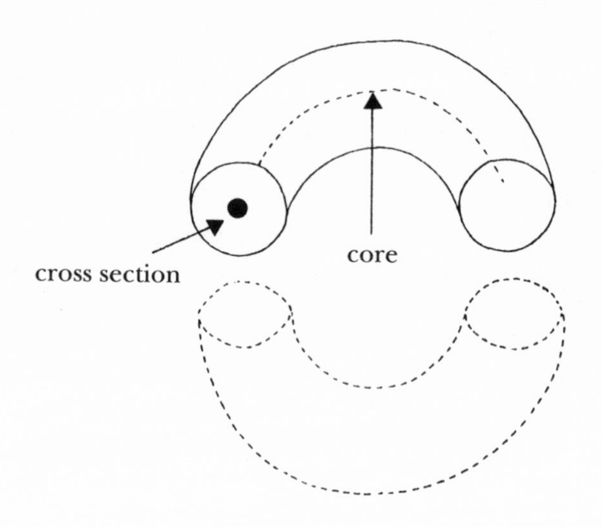

The torus is a solid constructed around a circle (the core of the torus), cross sections of which form disks. We will view only these cross sections from now on.

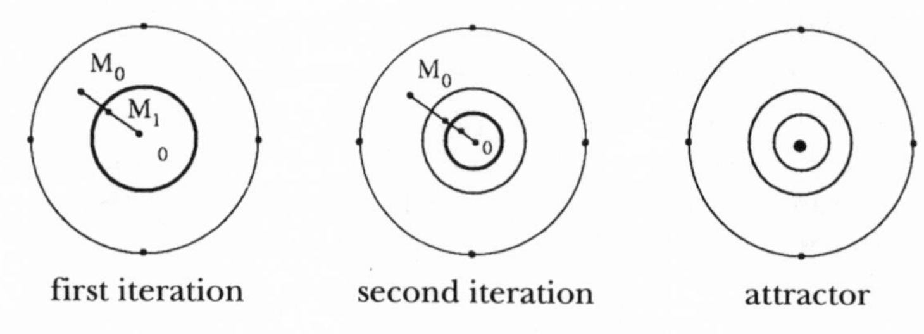

first iteration second iteration attractor

responsible for the example that we are about to describe, and which will open the door to the mysterious realm of strange attractors.

Let us get back to the torus we were just discussing, and contract it around its core as we just did. But this time we will have it undergo an additional transformation: we will stretch it and roll it onto itself. The result is no longer a torus, but a twisted ring that wraps twice around the core of the original torus, and which occupies a reduced volume.

We are interested in the way points move on the torus when this contracting and stretching is done over and over again. If M is one of them, we will call $f(M)$ the corresponding point after the preceding transformation; it is a point on the double ring, and thus a point of the torus since one is contained within the other. The evolution of the system is guided by the succession $M_{n+1} = f(M_n)$, which is the exact counterpart to the laws that we have encountered until now. But this time the behavior of the system is much more complicated. At each step, all of the positions are stretched in the longitudinal direction (the direction of the torus's core) and contracted in the two transverse directions (those of the cross sections). A first consequence of this is that the point M_n will no longer simply remain in a same transverse section, but will turn around the torus. But most important, the two effects combine to impose an attractor that is no longer either a point or a curve, but something much more complex which we will call a fractal, using Benoît Mandelbrot's terminology.

To see this attractor, let's try to follow the entire torus rather than one point. Its first transformation, obtained during the first stage, is the double twisted ring. At the following stage, each of these loops doubles and we obtain a quadruple twisted ring within the double twisted ring, itself within the initial torus (a simple ring). From stage to stage, we go from eight, to sixteen, to thirty-two twists, successively exhausting all the powers of 2. In doing so we obtain an infinite series of twisted rings, one inside the next, growing finer and finer, just as a sculptor's chisel works over a block of marble to bring to life a statue. But rather than some majestic or elaborate monument, in the end we have a sufficiently complex object to have earned the name *strange attractor*.

To get an idea of this, we can picture in our minds a braided ring, composed of an infinite number of rings around an initial torus. This notion is correct, of course, but doesn't quite convey the structural richness of a strange attractor. What is important, here, is the binary rhythm. We can easily represent the attractor at a given level of precision, by having a computer calculate the successive transformations of a given point for example. The choice of this initial point is immaterial; after a transitional phase, these transformations will be grouped around the attractor. If we look at their successive positions on a graph, we will see a cloud form which, as its contours become more and more distinct, will chart a network of capillary tubes coiling around the torus. But if for instance by increasing the precision of the calculations we view an enlargement, we will be surprised to see each of these capillaries split into two finer tubes. We had the impression that they were one only because of the imprecision of our observations. It is within this complex network that the points M_n, representative of the system state, circulate.

Let's try to understand this paradox. All the points of the torus represent potential states, and any one of them can serve as an initial state. But because of the system's natural evolution, after a brief transitory phase it confines itself to a much smaller, spiderlike region. Which means that most of these potential states will never be observed.

This is Smale's attractor, which may seem to the reader like a somewhat artificial construction. But meteorological equations also contain strange attractors, as Lorenz showed in a famous article. Lorenz's attractor doesn't have the exact same structure as Smale's; the latter was an intermediary object between a curve and a surface, whereas Lorenz's attractor is between a surface and a volume.

To give you an idea, let's imagine a book from which all the pages have been torn out except those pages with numbers containing only the digits 2, 3, 4, 6, 7, and 8 (no 0s, no 1s, no 5s, no 9s). For a one hundred page book, two distinct clusters of pages would remain, from pages 22 to 48, and from 62 to 88. These clusters contain only eighteen pages instead of twenty-seven; we would still have to pull out pages 25, 29, 30, 31, 35, 39, 40, 41, and 45. For a thousand-page book, obtained by subdividing each of the pages of the preceding book into ten thinner pages, we find gaps appearing in these homogeneous clusters. In the place of page 22 should be ten pages numbered 220 to 229, but four are missing: the pages 220, 221, 225, and 229. In fact, between pages 222 and 248 we find again a cluster of eighteen pages, in every way analogous to the one we observed at the previous stage.

We will notice that the proportion of remaining pages is multiplied by 6/10 at each stage; from 60% for a book of ten pages, to 36 for a one-hundred-page book and 21.6% for one thousand pages, which is to say that the book is thinning out. At the extreme, we must imagine a book with an infinite number of pages that become infinitely thinner, still within the initial binding. They've almost all been torn out, since the proportion of intact pages has fallen to zero, but there is still an infinite number, which are present in clusters of eighteen. More precisely, each page in this book is itself a book, a faithful reproduction of the original. And each page of these volumes, which appear in incalculable numbers, is itself a book built on the same model.

Lorenz's attractor looks like a surface folded onto itself in the usual three-dimensional space. But if we look at it through the microscope, we see a foliated structure appear, similar to the one we have just described. In other words, our instrument of observation, unable to embrace the whole of the at-

Smale's Attractor in the Torus

The deterministic law which is at the origin of Smale's attractor is the following:

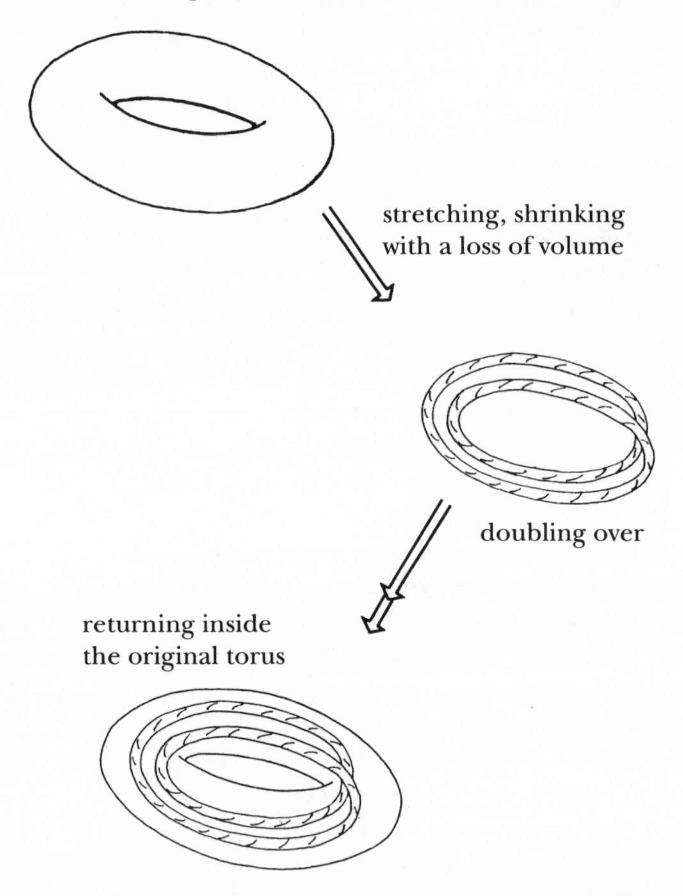

We thereby successively construct thinner and thinner tori, doubling over two, four, eight, sixteen, 2^n times within the original torus. Each is an approximation of Smale's attractor. The approximation is better the larger n becomes, that is, our power of resolution becomes higher. The attractor itself, which corresponds to an infinite resolution, is the limit. Thus, the gold necklace of Alleberg, made up, at first sight, of three tori juxtaposed, is on further inspection an interlacing of thinner and thinner tori accompanied by a fabulous prolifera-tion of fauna. The extraordinary techniques of the artist—filigree, granulation, embossing, impressing—merely allow him to create a precious yet imperfect approximation of the true object he carries within himself, and which, because of the sub-tlety and variety of its details, cannot be produced by the hand of man.

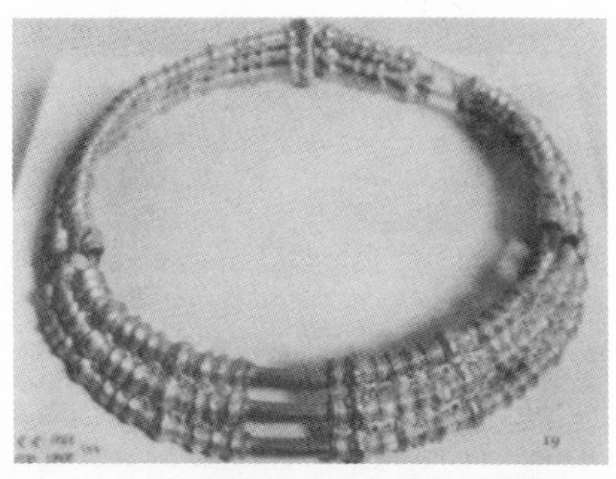

Lorenz's Attractor

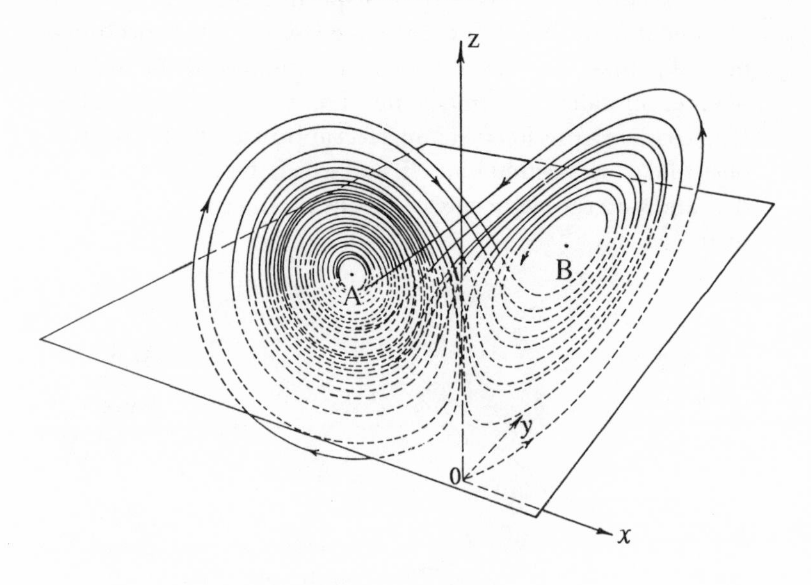

This three-dimensional drawing represents a typical trajectory in Lorenz's system.

$$\frac{dx}{dt} = -ax + ay$$

$$\frac{dy}{dt} = bx - y - xz$$

$$\frac{dz}{dt} = -cz + xy.$$

We see the trajectory, which begins at point *O*, form diverging loops around point *A*, then move on to point *B*, then return to *A*, then to *B*, and on and on indefinitely. We thereby observe an infinite succession of alternating oscillations that seem random, since the number of turns around a point each time is extremely variable. If we allow the movement to continue, the trajectory it describes builds into a foliated object, somewhere between a curve and a surface. This is Lorenz's attractor.

A very simplified representation of this movement can be seen on this carved stone from Vallsternarum. It represents a kind of labyrinth which from a distance suggests a clover with an articulation of each leaf. Lorenz's system contains an infinite number of leaves, all attached to the same stem. They are assembled into bouquets, each of which gives the illusion of being a single leaf. The complexity of the bouquet's underlying structure, however, is revealed by the chaotic behavior of the trajectories, which are forced to circulate within each leaf one after the other, and therefore to unpredictably flit from bouquet to bouquet.

tractor, artificially carves out pages in what is in fact a single, enormous sheet. Thus, if we study the movement of a specific point M_O, it will take a fairly long time to see the representative point M_n appear in our instrument's field of vision—and the greater the magnification used the more the field of vision will be reduced and the longer we will have to wait. If we continue this experiment, we must again wait some time for the representative point M_n to reappear in our field of vision so that we can see it again. We will then be observing it on another page of the book—it is outside of our necessarily limited field of vision that these pages touch one another and that the point M_n moves from one to the other.

Both Smale's and Lorenz's attractors share this property: if we examine them through increasingly greater magnifications we find the identical organized structure. We express this by saying that they are *fractals*. They are not like the usual figures of geometry, which are classified as one-dimensional, like lines or curves, two-dimensional, like planes or surfaces, and three-dimensional, like solids. Since fractals fall outside these classifications, mathematicians attribute to them intermediary dimensions. Smale's attractor is thicker than a curve and thinner than a surface, so its dimension is somewhere between one and two. Lorenz's attractor is thicker than a surface and thinner than a solid, so its dimension is somewhere between two and three. These intermediary dimensions show that the system doesn't occupy all of its space, or rather that only certain states are interesting from the point of view of dynamics. It is standard, in mechanics or physics, to introduce what is called the number of degrees of freedom: this is the number of parameters necessary to completely describe the state the system under consideration is in. It is 2 for Smale's system and 3 for Lorenz's. The fact that the attractor's dimensions are smaller means that the system will not explore all these possibilities, and will not exploit all the states theoretically possible (except during a brief transitional phase).

These ideas had the greatest success in hydrodynamics, because they have been useful in exploring one of the most important and least understood physical phenomena, namely, turbulence. The equations of Navier-Stokes, which regulate the movement of a fluid, have been known and studied for a long time. We know how to deduce the behavior of a fluid if the forces propelling it are weak, or if it is highly viscous. But once turbulence enters the picture, that is, when whirlpools appear, we are incapable of predicting the movement. It is accepted today that this situation is due to the presence of a strange attractor that appears above a critical threshold.

This hypothesis, which was stated for the first time in 1971 by David Ruelle and Floris Takens, is far from purely theoretical. It directly links experimental observations of physical instability with a mathematical instability underlying the equations of motion. Most important, it postulates that even though the system is of infinite dimensions (an infinite number of variables are needed to describe a single fluid state—its position and velocity must be indicated at each point in the volume it occupies), the turbulence occurs within finite dimensions. Indeed, the attractor is of finite dimensions, which we can calculate as a function of the conditions imposed on the fluid. The fluid evolves toward this attractor once the transitional phase has ended; the attractor will therefore be the site of the turbulence.

Once again we come to a brilliant intuition on the part of Andreï N. Kolmogorov, who, forty years earlier, had put forth the idea that the state of a turbulent fluid could be described by a finite number of degrees of freedom. This idea had come to him upon studying the degradation of energy in the course of movement, large whirlpools giving way to smaller whirlpools, until they become small enough to be dissipated into heat by the fluid's viscosity. Kolmogorov even calculated how many variables should be necessary to describe the state of a turbulent fluid; in other words, he had estimated the dimension of the attractor and, remarkably enough, his estimate has been confirmed by today's theories.

After this long digression, we can now return to the topic of entropy. For multidimensional systems, there can be both a loss and gain of information over time, corresponding to the simultaneous expansion in some directions and contraction in others. The loss of information is manifested by the fact that the system is quickly going to head toward an attractor, which may be much smaller than the entire set of possible states. It is on this attractor that most of the evolution will take place. The gain in information during this evolution is measured by entropy, as we indicated earlier. But this time, we measure it on the attractor. In other words, we begin by eliminating all redundant information, abandoning most of the available space in order to concentrate on the attractor. For instance, in the case of Lorenz's system, we study motion on the attractor only, neglecting points which lie outside, that is, the vast majority of points, because they will wind up on the attractor soon enough. For this restricted system, the remaining information becomes significant, and grows—or reveals itself—exponentially over time. Thus, if two points of the attractor cannot be distinguished due to the limited precision of our instruments, an additional observation may allow us to separate them. An entropy of 1 means that at each stage the number of points liable to be discerned, at a given level of precision, is multiplied by 10.

This definition has obvious flaws: we cannot count the "number of points." However small the segment of the attractor we consider, it contains an infinite number of points. Better to talk about "area" and "volume," with the understanding that the notions of area or volume discussed here are adapted to the attractor, notably to its own dimensions (which can be fractional). When we talk about area we are talking about two dimensions, talking about volume means three dimensions, which is why mathematicians prefer to speak about "measure," thereby avoiding any pronouncements concerning the dimensions of the attractor. To elaborate a satisfactory notion of "measure" for the attractor, a notion that can be adapted to the very particular structure of a strange attractor, has been one of the theory's main challenges. Today it has been largely resolved, thanks notably to the work of Yasha Sinaï, David Ruelle, and Rufus Bowen, who demonstrated the existence of an "ergodic measure" possessing remarkable properties. It allows us to reinterpret all of dynamics in statistical terms, and furnishes us with a probabilistic model for systems that can replace deterministic laws if the latter prove inadequate.

Let's elaborate on this point. Until now we have described the system from within; we have spoken about the number of degrees of freedom, that is, the number of variables which are required to specify the state the system is in, and we have used the deterministic law to construct an attractor in the set of possible states. But is this realistic? Outside of academic situations, the actual values of these variables are basically inaccessible to us. In hydrodynamics, for example, we are far from being able to measure the direction and velocity of the flow at each point of a fluid; yet this is what would be required to completely specify a state of the fluid. Similarly, the state of the atmosphere at any given moment, if it is understood as the pressure, the temperature, the velocity, and the chemical composition of air, measured around the earth and up to the interplanetary vacuum, is completely beyond our grasp, although it fully determines the weather. A possible alternative consists instead in proposing a variable considered significant, and in seeking to measure it as precisely as possible. In other words, we never observe the internal state M directly, that is, the myriad values of myriad variables necessary to fully specify a possible state of the system, but by the intermediary of a function $X(M)$. If the system begins with an initial state M_O, the succession of states $M_O, M_1, M_2, \ldots, M_n$ determines a series of values $X(M_0), X(M_1), X(M_2), \ldots, X(M_n)$ for the chosen variable X, and these are the values that we observe and must explain. We may be looking at the speed of the flow, measured at a particular point in the fluid, or at any other pertinent factor, like pressure. Of course, we can observe several variables simultaneously, like speed and pressure at a given point, which comes down to making $X(M)$ a vector with two components and not a number.

What Sinaï, Ruelle, and Bowen's results tell us is that a probabilistic explanation of the observations $X(M_O), X(M_1), X(M_2), \ldots, X(M_n)$ is not only possible but legitimate. The "ergodic measure" on the attractor is associated with a probability for this attractor, and the series $X(M_0), X(M_1), X(M_2), \ldots, X(M_n)$ has all the statistical properties of a sampling of n values taken independently and randomly on the attractor according to probability. For example, we may have recourse to the law of large numbers: when we add observations (n increases indefinitely), the empirical mean

$$\frac{X(M_0) + X(M_1) + X(M_2) + \ldots + X(M_n)}{n}$$

tends toward the mean value calculated according to this "ergodic probability" defined by Sinaï, Ruelle, and Bowen, with a probability of 1. From a statistical point of view, we can therefore consider that this series of values is random, and that it results from independent draws made on the attractor. It is experimental evidence for the "ergodic probability," and a way to compute the "ergodic mean" of any observable X.

The advantage of this probabilistic interpretation is that it is hardy; it requires no in-depth knowledge of the system, of its possible states, or of the laws that regulate its evolution. It is an all-purpose explanation, usable in many circumstances, even when, as in this case, the system is fundamentally deterministic. In other words, it doesn't matter if chance is in nature or if it is in the eye of the observer, the probabilistic explanation will not be affected. Note that it is not a matter of simply saying that this series of observations could be the result of random draws. It is true of any series of values: the essence of chance is that anything is possible, even the improbable. The probabilistic interpretation takes this one step farther: it states that the sampling observed is not only possible, but probable, from which it follows that the mean values, calculated over long periods of time, will end up stabilizing around their theoretical value.

This conclusion is general enough to remain valid even when we know almost nothing about the system. More precise knowledge would bring more precise information, not that a sampling will be selected according to a certain law of probability, with all of the inherent risks, including that of seeing our statistical predictions invalidated by chance, but what the exact results of the selections will be, like an experienced cheater who knows how to fool those around him. And like a gambler who wants to guard against fraud, the observer mustn't blindly refer to statistical methods. Only as a last resort can we allow ourselves to have recourse to a probabilistic method, when we have given up understanding the system from the inside and are ready to settle for a phenomenological description and statistical predictions.

Today there are methods that allow us to analyze a series of observations $x_0, x_1, \ldots, x_n$ in search of an underlying deterministic model. They consist essentially of interpreting each series of m values as coordinates of a point in m dimensions and in looking at the spatial distribution of the points obtained. Thus, if $m = 2$, we would construct the points:

$$M_0 = (x_0, x_1), M_1 = (x_1, x_2), \ldots, M_{n-1} = (x_{n-1}, x_n),$$

to study their distribution on the plane. If they cluster around objects of lower dimensions, forming curves or fractals, it is an indication that we are dealing with a deterministic system;

these objects are images of the attractor, which then has less than two dimensions. If on the contrary the points are distributed almost regularly across vast areas, forming uniformly gray patches, it is pointless to look for a deterministic model with an attractor of less than two dimensions. We must repeat the work in three dimensions, with the points

$$M_0 = (x_0, x_1, x_2), M_1 = (x_1, x_2, x_3), \ldots, M_{n-2} = (x_{n-2}, x_{n-1}, x_n).$$

Depending on how they are distributed in space, we may or may not detect the presence of an attractor of less than three dimensions. With a sufficiently powerful computer, and provided that we have enough observations, we can pursue this work up to $m = 10$, that is, we can detect attractors up to ten dimensions. In this way, attractors have been detected in turbulent flows and even in the stock prices on Wall Street.

Let's repeat one last time that while these methods allow us to recognize when there is a deterministic model underlying a phenomenon, thereby eliminating a form of chance that would be due solely to the observer's ignorance, a basic form of chance will still remain, if the underlying model has the property of exponential instability. This particular form of chance is measured by entropy. The best-informed observer will have instruments of only limited precision. Even if he knows the law of the system, he will know only the physical constants and the initial state to twelve decimals, but the thirteenth decimal whose value he doesn't know will gradually expand, eventually upsetting any predictions he might make and robbing them of all long-term validity. It is true that for deterministic systems, the initial state determines the final state. But exponential instability means that close knowledge of the initial state cannot produce close knowledge of the final state. Unless entropy is zero, the precision of the prediction becomes worse over time, meaning that continued observations are necessary to follow the system's evolution. The absence of observations leaves us somewhere on the attractor, in complete uncertainty.

The awareness of the difficulties raised by exponential instability in matters of forecasting dates back some time, although we may wish they had been more widely known. James Maxwell and Henri Poincaré, the greatest physicist and the greatest mathematician of the nineteenth century, wrote the definitive pages on the subject. Maxwell states repeatedly that while it is indisputable that the same causes produce the same effects, for certain systems that are sensitive to initial conditions similar causes will not produce similar effects. In *Science and Method,* Poincaré puts it this way:

A very small cause which escapes our notice determines a considerable effect that we cannot fail to see, and then we say that that effect is due to chance.

Developing this idea, he writes:

Why have meteorologists such difficulty in predicting the weather with any certainty? Why is it that showers and even storms seem to come by chance, so that many people think it quite natural to pray for rain or fine weather, though they would consider it ridiculous to ask for an eclipse by prayer? We see that great disturbances are generally produced in regions where the atmosphere is in unstable equilibrium. The meteorologists see very well that the equilibrium is unstable, that a cyclone will be formed somewhere, but exactly where they are not in a position to say; a tenth of a degree more or less at any given point, and the cyclone will burst here and not there, and extend its ravages over districts it would otherwise have spared. If they had been aware of this tenth of a degree, they could have known of it beforehand, but the observations were neither sufficiently comprehensive nor sufficiently precise, and that is the reason why it all seems due to the intervention of chance.[3]

3. Henri Poincaré, *Science and Method,* trans. Francis Maitland, pp. 67–68. (New York: Dover Publications, 1952).

This text is all the more remarkable as it was written in 1908, more than a half century before the discovery of Lorenz's attractor, at a time when we didn't yet have the power that today enables us to numerically simulate the behavior of the most general systems. It shows that Poincaré had a truly brilliant intuition, which he developed through rigorous work in the physical sciences, notably in celestial mechanics. This gave him occasion to study integrable systems closely and to develop an interest in the theory of perturbation, which at the time was the astronomers' main tool.

Before the advent of computers, the only way to study the dynamical system was to explicitly resolve the evolution equations, which is only possible for a very restricted class of systems, called integrable. For systems that are close to being integrable there are also methods that allow us to partially resolve the equations and to deduce the behavior of the system over large intervals of time, but not infinite ones. Thus, we can describe short-term behavior (as for all systems) and even medium-term behavior (during a length of time that is very difficult to estimate), but not long-term behavior (except if the system is integrable). This is what we call the method of perturbations, which is the basis of all astronomical calculations. We will have occasion to return to this later.

The notion of an integrable system has undergone several transformations over the centuries. It becomes fairly clear if we refer to the founding example, to the paradigm that inspired all modern science: Kepler's system, which describes the movement of a planet around the sun. This movement is entirely determined by three laws, which were discovered by Kepler after lengthy calculations: the path of the planet is an ellipse, and the sun lies at one of the focuses; it sweeps out equal areas in equal time (it therefore moves more quickly at the portions of the orbit that are close to the sun); its revolution period (planetary year) is proportional to the power $^3/_2$ of the large axis (thus a planet situated one hundred times farther away will revolve a thousand times more slowly).

Newton has been immortalized for having discovered gravitation, and for having solved the equation of gravitation in the very particular case in which the universe is reduced to two celestial bodies. In this way, he came upon Keplerian motion as a logical and necessary consequence of the law of gravitation $F = k/r^2$ (the force of attraction is inversely proportional to the square of the distance), and proved that this movement was perfectly and endlessly predictable. No matter how far we look into the future or retreat into the past, we can give the position of the planet. There is no trace of instability. Of course, if we were mistaken at the outset as to the planet's position or speed, we would perpetuate this error in calculating its course: the ellipse would be deformed or off track. But this calculation has been done once and for all. From now on, neither the calculated course nor the real course will budge, and the theoretical position must therefore remain close to the real position indefinitely. The errors will not increase over time, as they would in a case of exponential instability, and the entropy of the system is zero.

But the planet Earth is not the only one to revolve around its Sun. What we would like to know is whether equally desirable properties of stability extend to the solar system such as we know it, to large planets and their satellites, comets and asteroids, not to mention all as yet undiscovered celestial objects.

From the early days of celestial mechanics, this question arose in a very concrete manner. For example, because of the considerable influence of the sun's attraction, the orbit of the moon around the earth is far from elliptical, or even periodic—the exact description of its movement has occupied the greatest names in astronomy, from Newton to Poincaré. As for the orbit of the earth around the sun, it is much closer to the Keplerian ellipse. The main perturbation comes from the influence of Jupiter, which is on the order of $1/20{,}000$ of the solar attraction, whereas for the moon the relationship of the central force (terrestrial attraction) to the principal disturbance (solar attraction) is only $1/50$. In addition, the time scale is not the same; one terrestrial year equals twelve lunar months. As a result, we can make predictions much farther into the future for the earth moving around the sun than for the moon moving around the earth. But even for the earth there is a limit—which is difficult to pinpoint—beyond which the theory of perturbation ceases to be valid. We cannot know what happens beyond this horizon. We are therefore unequipped to answer a question as fundamental as this: Is the solar system stable? Will the earth's orbit remain close indefinitely to the one we now know? Or is it destined to escape into the interstellar void, or to be swallowed up by the sun?

Poincaré devoted a large part of his scientific activity to these questions. In particular he showed that the problem of three bodies, that is, the study of the movement of three masses in gravitational interaction, is not integrable. Thus, a simplified version of the solar system, reduced to the Sun, Jupiter and the Earth, is already beyond the point at which we can come up with a clear answer; the suspicion arises that the solar system is in fact chaotic. Of course, Poincaré didn't have the means to raise or to confirm this suspicion, though his work creates a strong presumption that would seem to confirm this notion. Even today the question is far from being resolved. And we have additional presumptive evidence, the result of numerical simulations performed by supercomputers used for astronomical calculations.

The most recent of these calculations simulates the evolution of the entire solar system over 200 million years, and brings to light an exponential instability—disturbances are multiplied by 10^{10} (10 billion) in 100 million years. That is, a fluctuation of one-tenth of a meter in the initial position can ultimately translate into a displacement of one million kilometers, and this over what is ultimately a short duration on the scale of astronomical or even geological time. On the other hand, over the first ten million years the computer indicates a great stability of the motion, which closely corresponds to the predictions of the theory of perturbations.

We mustn't suppose that a computer simulation of this type allows us to extend our ability to make predictions beyond this threshold of ten million years. Believing, for instance, that the positions which the computer gives for the planets in 200 million years are in any way accurate would be wrong. Once the system is unstable, the slightest disturbances take on importance, notably the slight inaccuracies that occur when the computer rounds off numbers, which it is forced to do at each step: it abbreviates every result of the interim calculations in order to shorten them to the desired number of decimals, and the cumulative effect of these small imprecisions can easily alter the final result. The computer is therefore unable to make long-range quantitative predictions. It creates strong presumptive evidence in favor of a qualitative result—that the system is chaotic—and that, over a duration on the order of several hundred million years, the planetary orbits may undergo major fluctuations.

Perhaps we should see this instability as the reason for the large climatic variations the earth has experienced over the course of its history. We already attribute the succession of ice ages, traces of which are found on the top geological layers of the planet, to small oscillations in the terrestrial orbit that last on the order of ten thousand years. But the time scale we are looking at here is much larger, and the potential effects devastating. From this point of view, we may wonder whether Mars enjoyed a more favorable position in the distant past, and whether Venus will take over for a blue planet ravaged by its inhabitants. Anything is possible. The movement of the planets, seen as a symbol of stability on the human scale, like a clock set by the Creator among the stars, is unpredictable and disorganized on the astronomical scale.

This example teaches us a great deal. First of all, it shows us that integrable systems, and thus predictable systems, are part of an extraordinarily limited category; Kepler's two-planet system is integrable, but by modifying it slightly (just adding a third, small, planet) we obtain systems whose entropy is no longer zero. Chaos is the rule, and predictability the exception. The study of celestial mechanics also saves us from a common ill—our desire always to find a cause. In a nonintegrable system, it is pointless to try to isolate causal filiations. If a demon displaces the earth a few centimeters from its orbit today, over a long enough period of time he will affect all the planetary orbits, and this effect can only be calculated, or even envisioned, by considering the solar system in its entirety. First, of course, the movement of the earth will change, and thus its perturbing influence on the other planets; in a second phase, once this effect is felt, the trajectories of these planets will slowly become modified. In the process, their respective positions will evolve, as will the relationships of their gravitational fields, and finally, on a scale of ten to one hundred million years, the entire solar system will be shaken.

As a result, we cannot calculate the effect of a single terrestrial palpitation if we limit ourselves to simply considering the earth's orbit; we do, of course, have to take into account the primary effect on this orbit, but also the secondary effect that changes in the earth's orbit will have on other planetary courses, because of Newton's law of gravitation, and the inevitable tertiary effect that the secondary effect will have once the changes in planetary paths have reverberated back onto the earth's motion, and so on. In other words, once we are out of the short term, we can deal with the system only in its entirety.

As a general rule, we cannot isolate a subsystem in a deterministic system; hence we cannot attribute a certain effect to a certain cause. If we introduce an initial shock, like our demon who alters the earth's course, the long-term consequences of this change can be evaluated only by retracing the complete evolution of the system with this new information. In the long range we will obtain a global state entirely different from the one that would have prevailed if the shock hadn't occurred, and it would be vain to seek to compare the two states. We cannot say that some things have changed and not others, and pretend to have identified the effects of the shock. A local modification can induce a global change, the way a wave of the magic wand can take us to a different world. The only effect we can attribute to this cause is the new situation in its entirety, which naturally provides little information.

The fundamental property of dynamical systems is that they can only be viewed globally. This rule suffers one exception: integrable systems, and foremost among them linear systems. Aside from these very particular situations, searching for causes or analyzing the consequences of a unique event leads us quickly into an unpredictable labyrinth, of which we remain prisoners unless we resign ourselves to the idea that no influence is too small to be neglected, and that the slightest palpitation throws into question the course of the universe. If this mode of thought is foreign to us, it is because we have the sensible habit of limiting our horizon to a future that experience tells us is foreseeable. Existing knowledge enables us to date the eclipses mentioned in ancient chronicles, and to predict tomorrow's weather. Over such short durations, the systems considered are approximately linear, and therefore integrable. It is only over the longer term that interactions due to nonlinear events complicate the picture, until the future of the most remote area becomes confused with that of the entire system.

To be even more precise about things, I would like to recall that a common definition of chance consists in seeing it as the intersection of independent causal sequences. A person is walking down the street just as a shingle comes loose from a roof; it hits him on the head and he dies on the impact. He was quietly going about his business, while the shingle was subject to the fancies of the wind. Two sequences of events, each with its own logic; they are so clearly separate, and their common result is so out of proportion, that we immediately bemoan his bad luck—a matter of chance.

Yet there is not, and there cannot be, independent causal sequences in the universe. By Newton's law of gravitation, the passerby on street level pulls on the shingle atop the building; the gust of wind that dislodged the shingle is inseparable from an entire meteorological framework in which the past activity of the victim had a role. To talk about independence is only a convenient approximation, a myopic view of events which we are forced to abandon if we are looking for more refined analyses or more distant horizons. A demon displaces an electron on Sirius, way below the range of our perception. In so doing, he modifies all the forces of attraction that this electron exerted on the other particles in the universe, notably on the gaseous molecules that constitute the earth's atmosphere. In just a few seconds this minute jolt, propagated and amplified by the collisions between molecules, translates into perceptible modifications. Meteorological instability then takes over, and the slight breeze that appeared in the Caribbean becomes a cyclone that devastates the eastern United States.

To try to isolate the causes of an event that has affected us is a necessarily limited venture; if pushed to an extreme, we might find ourselves investigating the movement of electrons on Sirius. We can only apprehend a small piece of the vast universe at a time, and we don't know when what we've forgotten is more important that what we see. We are like travelers lost in the fog; our gaze defines a small area that is reassuringly familiar, but beyond the gray walls that surround us the realm of the spirits begins.

It is the month of August in 413 B.C., and the Athenian expeditionary force is gathered before Syracuse. They have just suffered a crushing defeat while attempting to capture the hilltops overlooking the city, and their situation is now critical. Fortunately, they still have their fleet, which gives them the option of abandoning the siege to return to Athens or of finding a more favorable base of operations. Two Athenian generals, Demosthenes and Eurymedon, see the situation as urgent and want to embark immediately, but the indecision of the third, Nicias, delays the departure.

This leaves time for the Spartan Gylippus, who commands the Syracusan forces, to scour Sicily in search of reinforcements. In Selinus, he encounters a fleet sent to his aid by the Peloponnesians in the spring, and which is just now reaching its destination after a series of mishaps. Cast ashore on the coast of Libya by a storm, it found allied colonies there and stayed on to lend a hand against the Libyans, then traveled along the African coast until it was able to sail to Sicily by the shortest passage. It finally debarks at Selinus where it finds Gylippus, who immediately leads it to Syracuse along with other reinforcements.

The arrival of these reinforcements worsens the situation of the Athenians, who regret having missed their opportunity to slip away trouble-free. Since Nicias himself is no longer opposed, they begin making secret preparations for a departure. The troops are about to set sail in the dark, the fleet is about to leave this roadstead where it has been trapped and take to the high seas, when the lunar eclipse of August 27, 413 B.C. occurs. Overwhelmed by this incident, the men insist on waiting, the soothsayers declare that a delay of three times nine days is in order, and Nicias, a pious man, a believer in oracles and divination, won't even hear of leaving before that date.

Needless to say, the story ends badly for the Athenians.

Gylippus' reinforcements have enough time to train and gear up for the decisive battle. The Athenian fleet, squeezed into the Syracusan roadstead and unable to maneuver, is destroyed before the eyes of its troops. In turn these troops will be forced to attempt a retreat by land that should have been done by sea. Three days later the retreat is cut short by a debacle in which the generals and nearly the entire army perish, either by the Syracusan sword, before and after the battles, or in the stone quarries where they are detained in abominable conditions.

The immediate causes of this defeat, aside from Nicias's indecision, are two chance events: the arrival of the Peloponnesian fleet and the eclipse of the moon. Each one provides a perfect illustration of the intersection of two independent causal sequences, and therefore of a certain conception of chance. From the point of view of the Peloponnesians, the emergency fleet and Gylippus are part of one system, their actions are tending toward the same goal: victory over the Athenians. But from a certain moment on they lose contact; Gylippus in Sicily and the reinforcements in Libya are henceforth isolated subsystems acting independently, each following its own logic and reacting to the events it encounters. Each has an excellent reason to be in Selinus that day, Gylippus because he rationally planned his expedition there, the reinforcements because Selinus is the closest point to the African coast. These two reasons are independent, as are the adventures of each in Sicily or Africa that timed their arrival in Selinus. Chance lies in these two distinct entities meeting at the same place on the same day.

The case is more obvious for the eclipse of the moon. For we who know celestial mechanics and Kepler's laws, chance has nothing to do with the fact that an eclipse of the moon occurred the 27th of August in the year 413 B.C. The best proof of this is that the eclipse allows us to date the events related by Thucydides, and not vice versa. Astronomical tables give us the dates of eclipses for the entire historical age—three millennia counting back from today. We can also calculate the dates of eclipses for the next three thousand years: the respective movements of the earth, the moon and the sun are the outgrowth of a strict determinism and are perfectly predictable, at least on the scale of human history. These movements are obviously not influenced by the fact that a few thousand men are in the process of battling somewhere on the planet Earth. Conversely, taking into account the scientific knowledge of the era, neither the Athenians nor the Peloponnesians could imagine that lunar eclipses were predictable; they establish their strategies without taking this possibility into account. We therefore have a specific case in which two systems evolve independently, one following a physical determinism and the other a historical determinism, until August 27, 413 B.C., when the first affects the second in a major way.

For anyone embedded in historical determinism, there is nothing more troubling that the interference of an unexplained phenomenon, of pure contingency. Since the first anthropoid walked the earth, man has sought to survive by adapting to his environment, that is, by drawing lessons from experience to anticipate the future. Accepting the inexplicable, resigning oneself to the unpredictable, means leaving a gaping hole on the battlefield of our fight against nature's hostilities, and perhaps compromising the survival of the species. As a result, we feel obliged to attribute occult meanings to events that have no apparent meaning, that don't fit into the determinism we recognize after a thousand of years of experience. In such circumstances, primitive man will invoke the gods and try to appease them, while modern man will invoke chance and make statistical calculations. Only the discovery of a hidden determinism can bring the question to a close. As we have seen, this is far from easy, even for very simple systems; but as long as this explanation has not been found, no scientist can claim to be truly satisfied.

Viewed in this way, chance can only be relative to a human experience in a given historical situation. What seemed random or destined to Thucydides does not seem so to his modern reader. Today the eclipse that filled the Athenian army with superstitious fears would incite the usual curiosity that a rare and spectacular event generally arouses, and would be of nothing more than aesthetic interest. This brings our attention to the fact that chance is always a response to a question asked by man; if an event goes unnoticed, if it is considered uninteresting or explained in some other fashion, no one mentions chance. The eclipse of the moon raised this question for Nicias and his men, but today it no longer would. Coincidences as remarkable as the simultaneous arrival in Selinus of Gylippus and of the reinforcements sent to his aid happen every day, and no one notices them or is surprised.

Imagine that a Carthaginian merchant arrived in the port the same day as the Peloponnesian fleet. It is entirely possible, and even probable, given the intensity of commercial relations between the Carthaginians and the belligerent parties. But history would not have preserved this memory, since the event, even though it is just as remarkable in principle as Gylippus's arrival, would not have had the same significance. Of the three coincidences occasioned by this simultaneous arrival, Gylippus and the Carthaginian, the Carthaginian and the Peloponnesians, the Peloponnesians and Gylippus, only the latter attracts attention and raises questions. But they are all three equally remarkable or equally banal, according to one's point of view. They are equally banal, because there is nothing more common than the arrival of travelers at a port, and we would expect some of them to arrive on the same day. They are equally remarkable because they are equally improbable; all that is required for them to seem significant is an observer's willingness to draw them out of obscurity, the way King Midas turned everything he touched into gold. If we spoke Latin, we would say *egregium* for "remarkable," and the etymology would immediately translate our thought: *e-gregium,* outside of the flock, the way one individual can be singled out among many identical individuals and be marked by this arbitrary choice. The Carthaginian may be completely uninterested in the Peloponnesian war and the Sicilian expedition; he will be unaware of Gylippus's arrival, but he will be amazed if he runs into a long-lost friend among the newcomers. That's what strikes him, that's what he will remember and wonder about all his life, while the historian who relates the story of the war rises above individual destinies. For Thucydides, these many accidental coincidences constitute the background noise from which it is his task to distinguish the true signal, the one and only one worthy of being transmitted to future generations, the arrival of Gylippus.

Thus various points of view highlight one or another of the many events that occur simultaneously and that a perfectly detached observer would regard indulgently but with indifference, the way the smile of the Buddha greets our life's travails. The wheel of fortune is always turning and distributing destinies. Each one examines his lot anxiously, prompted by the certainty that he will only live once and by an awareness of his individual self. It is maya, the grand illusion. The Buddha sees the wheel turn, he contemplates the eternal cycle of reincarnations, he knows that the life allotted me today is just an episode in an infinite history in which I will play every role, one after the other. There is no chance because there is no meaning, no reason to highlight a particular moment in this history, or in this history rather than another. The inner quest that makes us cry, "Why is this happening to me?" can lead only to non-sense and suffering. Chance dissolves in the sweet indifference of the world.

5

Risk

After so many digressions, the reader will pardon us for leaving the work of Snorri Sturluson for a moment to venture into *The Story of Burnt Njal.* It is the latest in date of the great Icelandic sagas, but also the longest and perhaps most successful, a final and magnificent blossoming of a genre that was to fade soon after into the court romance and the chanson de geste. The story of Burnt Njal is still based on strictly historical facts—the central episode, the fire at the manor of Bergthorskvall in which Njal and his son perished, is confirmed by other sources including the Landnamabok, as is the pitched battle that took place at the althing (supreme assembly) of 1011 during the trial over the arson. The anonymous author of the saga preserved the concise style already characteristic of Snorri; the entire work is a monument to a civilization on the verge of annihilation. Njal sees the events and the final destruction of his family approaching, just as Odin foresees the Ragnarok, the revolt of the giants and the end of the world, but is incapable of preventing it.

The friendship between Njal of Bergthorskvall and Gunnar of Hlidárendi is one of the keys to the work. Gunnar is a hero able to jump his own height in full gear, and a peerless archer. He trusts Njal in all matters and seeks only to live in peace, but neither the wise advice of one nor the good intentions of the other can prevent an inevitable chain of aggression, retaliation, and vengeance from driving Gunnar to his end. When the episode begins, Gunnar has submitted once again to the verdict of the althing, which condemns him to three years in exile. There is nothing dishonorable about this; his social position will even be strengthened by the added glory and possessions with which he will inevitably return from his foreign expedition. On the other hand, if he doesn't leave, he will be breaking the law, and thus at the mercy of his enemies, who will be free to kill him without punishment. Here is the account of Gunnar's departure from Hlidárendi. His brother Kolskegg is accompanying him.

Gunnar made men bear down the wares of his brother and himself to the ship, and when all Gunnar's baggage had come down, and the ship was all but ready, then Gunnar rides to Bergthorskvall, and to other homesteads to see men, and thanked them all for the help they had given him.

The day after, he gets ready for his journey to the ship, and tells all his people that he is riding away for good and all, and men took that much to heart, but still they said that they looked forward to his coming back afterwards.

Gunnar threw his arms round each of the household when he was ready, and every one of them went out of doors with him; he leans on the butt of his spear and leaps into the saddle, and he and Kolskegg ride away.

They ride down along Markfleet, and just then Gunnar's horse tripped and threw him off. He turned with his face up towards the hillside and the homestead at Hlidárendi, and said, "Fair is the hillside; fairer that it has ever seemed to me; the corn fields are white to harvest, and the home mead is mown; and now I will ride back home, and not fare abroad at all."

"Do not this joy to thy foes," says Kolskegg, "by breaking thy

atonement, for no man could think thou wouldst do thus, and thou mayst be sure that all will happen as Njal has said."

"I will not go away any whither," said Gunnar, "and so I would thou shouldest do too."

"That shall not be," says Kolskegg; "I will never do a base thing in this, nor in any thing else which is left to my good faith; and this is that one thing that could tear us asunder; but tell this to my kinsmen and to my mother, that I never mean to see Iceland again, for I shall soon learn that thou are dead, brother, and then there will be nothing left to bring me back."

So they parted there and then. Gunnar rides home to Lithend, but Kolskegg rides to the ship, and goes abroad.

Hallgerda was glad to see Gunnar when he came home, but his mother said little or nothing.[1]

"Hallgerda was glad to see Gunnar when he came home, but his mother said little or nothing." Rarely has anyone gone this far in the art of understatement and the expression of disapproval. Hallgerda is Gunnar's wife, and the reader of the saga has had plenty of time to become familiar with the vindictiveness and animosity she bears toward her husband. It is not because she loves him that she is happy about Gunnar's return, but because she now has a chance to get revenge. Gunnar's mother, on the other hand, who just bade farewell to her two sons without being sure she would see them again before her death, becomes silent when one of them returns. She knows, he knows, and everyone knows that it's suicide. Hallgerda's joy is obscene, as would be any reproach from his mother; it would be pointless as well, since Gunnar knows what she is thinking. She tends to her chores, painfully silent.

Gunnar's decision is as sudden and irreversible as a natural disaster. There is no warning. He is present at the althing

1. *The Story of Burnt Njal,* trans. Sir George Webbe Dasent, chap. 43 (London: J. M. Dent, 1911).

when his case comes up, he witnesses Njal's efforts to arbitrate, and displays no displeasure at the outcome. On the contrary, he promises Njal to honor it and prepares for his departure. He says goodbye to his family, and it is only by accident, because of a slippery stone or a passing shadow, that his horse stumbles and offers him an occasion to change his mind. This turnabout occurs at the worst moment. It causes his brother Kolskegg to leave him behind, and is perhaps the only act that could discourage such an indestructible ally. From this moment on, the events accelerate. Gunnar is condemned at the althing the following summer; in the fall his enemies mount an expedition that finally does him in.

All this because, by chance, he turned around; his gaze settled on the farm where he'd been living and it was as if he were seeing it for the first time, nestled in the hill, surrounded by luminous, fragrant fields. Why did the horse stumble? Do such momentous decisions really depend on incidental circumstances? If the hoof had come down ten centimeters further, or if the procession had passed ten minutes later, this fall wouldn't have happened, or Gunnar wouldn't have been the one to fall. He and his brother would have completed their three years of exile, returned to their country with glory and honor, their enemies definitively silenced. The siege of Hlidárendi, in which Gunnar perished, and the fire of Bergthorskvall, which consumed Njal and his sons, wouldn't have occurred, and we wouldn't be reading the saga of Burnt Njal today.

Gunnar says his goodbyes to his mother, his wife and sons, and his friends. He rides toward the shore side by side with his brother; in a few seconds they'll be at the sea, in a few hours the island will disappear from the horizon. His thoughts are already miles ahead in the Orcades or in Norway. What destiny awaits him there? A chance occurrence, a glance, and Gunnar changes his mind and his future. He cannot defend his decision and he knows it; it is irrevocable, and he will stick to it. His friends do their best to dissuade him; all he can do now is refuse their help, for fear of bringing them down with him. All this because the country is beautiful, at dawn, near Hlidárendi.

According to decision theory, each person envisages every possible event and assigns it a probability that translates its relative plausibility. Zero probability signifies that the event in question is considered to have no chance of occurring, and can therefore be disregarded. An event with a probability of 1, on the other hand, is considered to be certain; in other words, we are sure that it will happen. The intermediary probabilities, between 0 and 1, indicate various degrees of certainty, just as the gradations of the thermometer between 0 and 100 Celsius measure the temperature of water.

We can arrive at these probabilities in several ways. The most natural one consists of calling in experts. This is how industrial risks are generally evaluated, and how events like the "China syndrome," the melting down of the core of a nuclear reactor, are assigned an infinitesimally small, but nonetheless positive, probability—between 10^{-10} and 10^{-5} depending on the authors of the study and for whom it is designed. In an analogous fashion, we calculate the probability of having a car accident at a given moment on a given route, or the probability of a space-shuttle launch failing.

The idea behind these evaluations is that a major event like an accident is never more than the catastrophic result of a combination of minor circumstances, a web of small coincidences that individually would have been unimportant but that unfortunately accumulate to trigger phenomena of another order. Several years ago, the nuclear reactor at Three Mile Island was the site of a major accident because a valve didn't close when the light on the control panel indicated it was closed. As a result, the operators worked for several hours with a false image of the situation, and the measures they took made it considerably worse. This is a scenario which, with just one more piece of bad luck—another valve opening or a conduit breaking—could have led to an even more serious accident, even to the famous China syndrome. The probability of a valve not closing can be reasonably evaluated by an engineer, as can the probability of a control indicator not lighting up; it is therefore possible to calculate the global probability of such a scenario. By making an exhaustive catalog of all the scenarios that could lead to the China syndrome, and by evaluating the probability of each one, we can establish a probability for the China syndrome itself. This then becomes a management instrument, in that it allows us to objectively calculate the risk. Any technological improvement that diminishes the probability diminishes this risk. Nevertheless it may not always be beneficial, since it may increase other risks, like the risk of polluting the atmosphere, which is calculated in a similar fashion.

We may be surprised to see probabilities assigned to events that have never happened, and which we hope never will. The idea is that the events in question break down into independent micro-elements that must occur simultaneously, the way opening certain doors requires the presence of three people, each with a different key. If these people pass by the door once in every ten days on average, and if the dates of their passage are independent of one another, the probability that we can open the door on any given day is $1/10 \times 1/10 \times 1/10 = 1/1{,}000$, that is, once every three years on average. If behind the door are the four horsemen of the Apocalypse, we may find the risk of their destroying the earth once every three years to be too high. So we put eight locks on the door instead of three. The same calculation gives a probability of one in a hundred million, about once every 3,000 centuries. Since human history only dates back about fifty or sixty centuries, we will feel confident that the end of the world has been reasonably delayed.

But decision theory doesn't require us to have objective bases for the probabilities we assign future events. I may be perfectly convinced that the world will end tomorrow morning. I will assign this event a probability of 1, and act accordingly. This is a subjective probability; it is valid not by the force of my reasoning or by the number of people who agree with me, but because of my intimate conviction. If nothing happens tomorrow, I will be forced to revise this probability; in the meantime however it will inform my actions. And indeed, every thousand years or so millenarians go into gear, predicting that the end of the world is near and calling on their followers to abandon worldly goods and prepare for the return of the Lord. The fact that a conviction is irrational doesn't diminish it; the fact that a probability is subjective doesn't mean it won't influence a decision.

I can therefore evaluate the probability of events about which I have only sketchy information. Who hasn't heard barroom conversations in which the guilty parties of unsolved crimes are identified, or the secret illnesses and impending deaths of prominent politicians revealed? The people who spread these rumors must believe them to some degree, which means they must take them into account in making decisions. I can even assign a probability to an event which I know nothing about. If I have to bet on a game of badminton and all I know are the names of the players, I'll flip a coin, that is, I'll assign each player an equal probability of winning. If it's tennis, there's a chance the names will mean something to me, in which case I can abandon this 50%–50% strategy, which merely reflects my lack of information. Taking this one step further, I may be asked to evaluate the force of my convictions by establishing the degree of confidence I have in my subjective probabilities. If I am sure of myself, if for example the players are far apart in the ATP ranking, I will say that my evaluation is probably correct, and I may even go so far as to calculate this probability. In this way we arrive at complicated responses such as: "I think that X has a 90% chance of beating Y, and given the information at my disposal I am 60% sure of what I'm saying."

It is clear that probabilistic language is being used to express two different situations here, and that its precision is better adapted to tennis than to badminton. When I say that X has a 90% chance of beating Y, I am referring to a mathematical model in which the random elements that will determine the outcome of the match have been identified and weighed. The figure 0.9 is the result of what may be a cursory but nonetheless necessary calculation. In an ideal situation I would be equally familiar with X and Y, and would have all the elements necessary for an evaluation at my disposal. If this were the case, I would no doubt arrive at a fairly precise estimate of each one's chances. At the other extreme, if I know neither X nor Y, if I've never even heard of either one of them, I have no information to steer me one way or another. This is what I will call, using the terminology of many psychologists who have worked on the subject, a situation of "ignorance"; the preceding situation, for which we have an exact probabilistic model, would be qualified as "random."

In practice we often find ourselves in intermediate situations, between these two extremes. We speak then of "uncertain" situations. This is what the forecaster was expressing when he said he was 60% sure of his estimate. He was closer to randomness than to ignorance, the two extremes of the scale. We can recreate the transition from randomness to ignorance by organizing lotteries. Two encounters are organized, the first between two players I know perfectly well (a purely random situation; I am 100% sure of my model), the other between two players I know nothing about (situation of total ignorance; no confidence in my model). We are going to draw lots on which one will take place. At this point I am in a situation of uncertainty, and my level of uncertainty depends on the probabilities chosen. If the first match has six chances out of ten coming up, I will say that I am 60% sure of my model, or that my level of confidence is 60%.

It is therefore possible to pass continuously from a purely random situation to total ignorance, through every degree of uncertainty, and to calculate these by probabilities. The making of a decision in an uncertain future can thereby—at least in principle—be reduced entirely to evaluating probabilities. The classic example is the Pascalian wager, such as it has been understood and perpetuated by the authors of textbooks. The probability that God exists is slim (perhaps 10^{-10}), but the advantages we gain from his existence are so great (an eternity of happiness, perhaps 10^{1000} in unspecified units), that the gain ($10^{-10} \times 10^{1000} = 10^{990}$, quite a respectable sum) should inspire us to act as if God existed. The other side of the wager, which has a greater probability ($1 - 10^{-10}$, let's say 1 so as not to have to count the decimals), but with a far weaker gain (at best 100 years of happiness, or 10^2 in the same units as before), leads to a potential gain of $1 \times 10^2 = 10^2$, far inferior to the first. To be fair to Pascal, this was most likely not exactly what he meant.[2]

Unfortunately things are not so simple, because human beings commonly display an aversion to risk. In the situation we just described, the potential gain is certainly immense, but the probability of success is so weak that in the end the wager seems unattractive, taking into account the magnitude of the initial investment (probably a lifetime of renouncement and asceticism). We're happy to bet $1 at a hundred to one. We would hesitate to bet $1,000,000 under the same conditions. This type of aversion to risk is well known to economists and financiers. The proverbial bird in the hand is worth two in the bush. This is why an investment considered to be risky, a junk bond or a hi-tech company, must offer a higher return than

2. Pascal, *Pensées,* ed. Jacques Chevalier, in *Oeuvres complètes* (Paris: Gallimard), in particular the *Discours* by Filleau de la Chaise (pp. 1474–1501).

investments considered safe, or simply less risky, in order to coexist on the same financial markets. A "risk premium" is subtracted from the anticipated return, and the larger the perceived risk, the higher this premium.

But added to this aversion to risk is another aversion, to uncertainty, which is different in nature and far more difficult to calculate. If two equal players face off, and I know them to be equal, I will assign each one a 50% chance of winning. If the match is between two unknowns, I will again assign each one an equal chance of winning, but I will be in a more uncomfortable situation. In fact, as a famous psychological experiment performed by D. Ellsberg[3] shows, the application of probabilistic formalism to situations of uncertainty leads to strange paradoxes.

Ellsberg's experiment is the following. Subjects are presented with urns, each containing 100 balls. They are told that the first contains exactly 50 red balls and 50 black balls; the second, they are told, contains either black and/or red balls, but the proportion of each is left unspecified.

The betting starts on the first urn. The subjects choose a color and then a ball is drawn. Those who guess correctly win $100, the others nothing. The experiment shows that most people bet indifferently on the red or on the black, that is, they assign them subjective probabilities of 0.5 and 0.5.

Next comes a series of bets on the second urn. The conditions of the betting are the same, but this time the subjects have no information as to the contents of the urn—except for the fact that the ball will be either black or red. This is a situation of ignorance, while the previous one was an aleatory situation. In keeping with the theory, most people bet indif-

3. D. Ellsberg, "Risk, Ambiguity, and the Savage Axioms," *Quarterly Journal of Economics* 75 (1961): 643–69; and *Reply*, 77 (1963): 336–42. See also, H. Einhor and R. Hogarth, "Decision Making under Ambiguity," in *Rational Choice*, Hogarth and Reder, eds. (Chicago: University of Chicago Press, 1986, pp. 41–66).

ferently on the red or on the black, that is to say that they still assign them subjective probabilities of 0.5 and 0.5.

Finally we come to the paradox. We open the third round of betting. Participants win $100 for a red ball and nothing for a black ball, but they have the right to choose the urn from which the ball will be selected. The fact that the subjective probabilities are the same for the two urns would lead us to assume that the subjects consider that they have a one in two chance of winning with either urn. According to the theory, then, the subjects should choose indifferently between one or the other. But they don't: most people express a distinct preference for the first urn (the one with known proportions). This preference is even more pronounced if it is a matter of losing $100 rather than winning—but we find fewer volunteers. It is therefore as if ignorance were a supplementary risk factor which subjective probabilities alone don't take into account.

Two questions arise from this analysis. What is risk and how can we control it? We make a distinction between randomness risk, a natural outgrowth of the usual methods of probability theory, and ignorance risk, due to an absence of information about what can happen. The distinction would not stand up to close analysis (is chance anything more than the expression of our ignorance?), but it has great practical importance. Experience shows that human beings more readily accept randomness risk than ignorance risk, even if in most concrete situations the two risks are both present. Popular wisdom tells us "better the devil you know than the one you don't know," and the Bard too tells us this in the mouth of Hamlet:

> For who would bear the whips and scorns of time,
> The oppressor's wrong, the proud man's contumely,
> The pangs of despis'd love, the law's delay,
> The insolence of office, and the spurns

> That patient merit of the unworthy takes,
> When he himself might his quietus make
> With a bare bodkin? Who would fardels bear,
> To grunt and sweat under a weary life,
> But that the dread of something after death,—
> The undiscover'd country, from whose bourn
> No traveller returns,—puzzles the will,
> And makes us rather bear those ills we have
> Than fly to others that we know not of?[4]

Indeed, every day individuals and nations confront what are sometimes major risks, the limits of which have been established through experience, but facing the unknown is quite another story. Human life is composed of accepted risks. Primitive man, who lives from hunting and gathering, doesn't know if he'll find anything to eat tomorrow. The peasant sows today not knowing if the harvest will reward his efforts in six months or a year. The businessman who replenishes his stock doesn't know if it will sell. Each of these risks is within the realm of a tested model, passed down from generation to generation, which assigns limits to possible events and determines certain probabilities. The peasant knows from experience or hearsay the various disasters that can attack his harvest: frost and drought, flood and fire, insects and disease. And he can imagine worse, the ten plagues of Egypt, for example, even though they have never occurred again, to man's knowledge. But he knows that the sky won't fall on his head. He knows that all these disasters have small probabilities, which can be reduced by taking certain precautions and practicing certain rituals, as is proven by the simple fact that after so many generations he is still here cultivating the same earth. He must have a reasonable chance of survival, as his ancestors did before him, and if the year turns out to be a bad one the next year will be better.

4. Shakespeare, *Hamlet,* Act III, Scene I.

Now let's imagine the conquistadors storming the empire of the Incas. This society was accustomed to taking a certain number of risks. It was a peasant society, accustomed to agricultural risks; it was also an empire established by conquest, so military risks were not unknown either. Yet it seems they didn't know how to confront the new risk that these bearded and armored creatures presented, riding strange animals and bearing thundering lances, whose manner of fighting defied the ordinary laws of humanity. The reasons for this society's collapse disappeared with Atahualpa and millions of his subjects, but it seems plausible that it was linked to a psychological inability to assume certain risks. Better to give in than to fight an adversary whose power couldn't be measured.

We generally don't think of peasants or merchants, people who take certain kinds of risks every day, as needing tremendous courage. The character of the explorer, on the other hand, is the archetype of courage. Why? Because he penetrates into this *terra incognita,* into this area shown in white on maps, the sight of which makes us so uneasy that cartographers prefer to cover such an area with illustrations, or to surround it with an inscription: *Hic sunt leones.* This type of information, even if it is fantasy and recognized as such, is more reassuring than no information at all. And yet, this white spot on the map may be the Eldorado, where every river is full of gold nuggets, or the mythical land that the Lord has promised to his people, the land of milk and honey. The explorer might be sailing into great fortune; why do we always imagine the worst, if not that we fear the unknown?

So it is that we find two areas of risk becoming distinct, the realm of the randomness, characterized by probability theory, and the realm of the unknown, in which the only rule is caution. This would no doubt be enough to give a coherent image of human decision, if the border between these areas weren't hazy, and if probability theory hadn't gradually become a universal instrument. In evaluating the risks corresponding to nuclear power plants, for instance, the political and technical communities, relying on probabilistic models, are at odds with public opinion, which has a tendency to be far more prudent. Is it simply a matter of a lack of information or expertise on the part of the public, or is the validity of probabilistic models in question? The importance of this problem justifies devoting a few pages to it.

The first thing to note is that all too often when official estimates of the probabilities of an incident are made public, they turn out to be in blatant contradiction with the observed frequencies. The most famous example is from NASA, which in 1985 estimated the probability of an accident occurring to the space shuttle to be 1 in 100,000,[5] whereas previous studies had concluded there was a probability on the order of 1 in 100, and in fact the carrier rocket exploded on the twenty-fifth launching. I don't know the official estimates, if they were even published, but I doubt that the probability of accidents at Three Mile Island or at Chernobyl had been estimated at their true figures. Finally, let's remember that the *Titanic* couldn't sink; the probability was not even infinitesimal, it was zero. This fact was widely publicized, yet the ship sank on her maiden voyage.

We may always suspect shady politicians or fanatical engineers of playing with the figures to get their projects through, but even in circumstances in which there are no material interests at stake, the probabilities of failure calculated by competent people in good faith may turn out to be systematically underrated. A recent article examines twenty-seven measurements of the speed of light, published between 1875 and 1958.[6] As is customary, each of them was accompanied by an estimate of the margin of error offered by the researcher

5. *Space Shuttle Data for Planetary Mission RTG Safety Analysis* (Ala.: NASA, Marshall Space Flight Center, 1985).

6. M. Henrion and B. Fischhoff, in *American Journal of Physics,* 54, 1986, p. 791.

himself: he provides us with a so-called "standard error" which, for every quantity ε, allows us to calculate the probability that the value given for the speed of light deviates from the true value by more than ε. If we now take for ε the gap between the measured value and the 1984 value, which is universally accepted nowadays as the true one, in each case we find probabilities of less than .5%. In other words, if we accept that the successive researchers have been right in estimating their chance of failure, we can only conclude that the events actually banded together against them to invalidate their results: twenty-seven times in a row they achieved a result that had less than five chances in a thousand of occurring.

The second observation is that unexpected things do happen, often as a result of human error. In 1987, two scrap workers in the city of Goiânia, Brazil, found in an abandoned clinic a capsule containing about 100 grams of a phosphorescent powder. It turned out to be radioactive cesium 137, to which an entire population wound up being unexpectedly exposed. By the time the spread was finally checked, in December 1987, 121 cases of contamination had already been reported, four of which were fatal, and more than 100,000 inhabitants showed up for radioactivity examinations. On foreign markets, the fear of contamination drove down the price of agricultural products from the state of Goiás to half its usual level, and industrial production suffered equally. In short, there was a considerable human and economic price for a risk that no one was even aware of running. And the same discovery could have had more catastrophic results. What would have happened, for example, if the cesium 137 had fallen into the hands of blackmailers, and if Brazil had been threatened with an organized contamination?

Experience proves that when it comes to anything nuclear, the risk defies measurement. Unexpected scenarios led to the accidents at Three Mile Island and at Chernobyl, aided by the inadvertent cooperation of the power station at each location. We can talk about human error, but that is an easy way of shifting responsibility from the designers of power stations to the operators. Human error should be integrated into the security mechanisms and reliability calculations, just as technical errors are. From this point of view, an accident due to an operator's error is far more serious than a mechanical breakdown, for if error or negligence can be so dangerous, what about intentional destructiveness? We may think we are eliminating these risks by designing automatic power stations. But the human risk factor does not lie solely with the operators. After all, engineers can make mistakes, experts can lie, guards can fall asleep. The people who make decisions—and the public—owe it to themselves to take all these risks into consideration, and they are in a better position than the technicians to appreciate them.

We have to consider that a single neglected or unrecognized risk can invalidate all the reliability calculations, which are based on known risks. In addition, the factors that aggravate the risk largely outweigh the factors that would tend to diminish it, just as bad money drives out the good. Imagine for example a power station where the probability of an accident is estimated at one in a million. Imagine that two risk factors had been forgotten, one multiplying the risk by 1,000 and the other diminishing it in the same proportion, so that during 10% of the time the power station in fact functions with a $1/1{,}000$ probability of an accident, and during 10% of the time with a $1/1{,}000{,}000{,}000$ probability. During the remaining 80% the previously estimated probability, $1/1{,}000{,}000$, is valid. A simple calculation then shows that the real probability of an accident is on the order of $1/10{,}000$. This probability remains virtually unaltered by measures taken to diminish the known risks. If strict and assiduous work on the security standards reduces the probability of an accident during the 90% of the time when the unrecognized risk isn't operating to

$1/1{,}000{,}000{,}000$ (one in a billion instead of one in a million), the real probability of an accident will still be about $1/10{,}000$.

To these discouraging considerations we must add yet another. It is that never in the history of humanity have we made decisions that reach so far into the future. The nuclear industry produces waste that will remain extremely dangerous for at least ten thousand years. The wastes that aren't reprocessed or lost (this happens) are warehoused in special sites, abandoned mines or granitic caves, where in principle they are under constant surveillance. But ten thousand years is twice the age of writing, twice the duration of human history. Imagine if distant ancestors in the mists of time, long before the first Egyptian or Chinese dynasties, long before the emergence of the religions known today, had left behind crypts not to be opened or even approached. Would the guard remain diligent during the long succession of empires, wars, and disruptions? Would the command be passed down, would the memory stay alive? Or would some bold conqueror order them opened in his presence, to confront the legend?

We comfort ourselves with the hope that nuclear wastes will not remain warehoused for ten thousand years, that long before the end of that period future generations will have found a cure for cancer, a vaccination for AIDS, the secret of eternal youth, a way to dispose of our garbage. The waste will be reprocessed in nonpolluting factories, or sent into space, aboard missiles that will not explode in flight. These same missiles will no doubt be useful for rebuilding the ozone layer, and we eagerly await our descendants' solution for diminishing the level of carbon gases in the atmosphere.

The truth is that industrial civilization moves forward without measuring the risks incurred, and without thinking globally. It is easy to construct an indictment of nuclear energy, but burning fossil fuels isn't good either, and even hydroelectric dams have disadvantages. We need to address the issue of energy sources for the planet, as well as so many other issues that don't receive sufficient attention. When will the AIDS epidemic end? Already in certain African countries a third of the population is infected. Have we considered the consequences? What about the fact that entire populations have been driven from their land and have been living in refugee camps for more than forty years? By allowing this type of situation to perpetuate, generation after generation, we deliberately create major historical risks. Who measures them? Who takes them into account?

We move forward, ignoring the risks we create. From time to time an accident shakes us from our numbness, and we glance into the precipice. An accident at Chernobyl, and we no longer drink milk; the death of an actor, and the police wear gloves; a popular uprising, and we'll vacation elsewhere. We are like Gunnar of Hlidárendi, who comes to himself after falling from his horse. His response, however, had a different sort of motivation and dignity.

Taking a risk is not always the result of a calculation—far from it. Already in economic affairs, as Keynes pointed out, "If human nature felt no temptation to take a chance, no satisfaction (profit apart) in constructing a factory, railway, a mine or a farm, there might not be much investment merely as a result of cold calculation."[7] A little further back he declared that "when enterprises were mainly owned by those who undertook them or by their friends and associates, investment depended on a sufficient supply of individuals of sanguine temperament and constructive impulses who embarked on business as a way of life, not really relying on a precise calculation of prospective profit."

Why did the Athenians take the risk, at Marathon and at

7. *The General Theory of Employment, Interest and Money,* chap. 12, III.

Salamis, of confronting an enemy with countless troops, when so many other Greek cities had capitulated, and while the Lacedaemonians were retreating to the Peloponnesus? A half century after the events, listen to the pride with which the Athenian ambassadors spoke to the Lacedaemonians in Sparta: "This is our record. At Marathon we stood out against the Persians and faced them single-handed. In the later invasion, when we were unable to meet the enemy on land, we and all our people took to our ships, and joined in the battle at Salamis."[8] The risk was taken twice. If the undertaking had failed, no one would ever have heard of the Athenians again. But it succeeded, and we're still talking about them, twenty-five centuries later. Aeschylus, the author of so many masterpieces, wanted no other claim to glory on his tomb than that of having fought at Marathon: "In the fertile Gela, in this tomb lies Aeschylus, son of Euphorion, an Athenian. He is dead. But the Mede with long hair experienced the glorious site of Marathon, which speaks of his valor."

These are cases in which behavior seems dictated not by computed probabilities but by moral imperatives which, once accepted, leave little choice. The soldier engages in desperate combat because of a feeling of honor, or out of solidarity with his companions. The capitalist, driven by the Protestant ethic, feels that money must bear fruit, as in the parable of the talents (Luke 19: 12–25). The moment one's conscience draws up rules for decision-making, without considering the alternatives, the very notion of risk vanishes and is replaced by the idea of fate. When Gunnar falls from his horse and decides to stay, he doesn't weigh the pros and cons; fate imposes itself. He won't travel the seas, exiled by his enemies' conspiracy, even though he'd be able to return in three

8. Thucydides, *History of the Peloponnesian War,* Book I, chap. 73, trans. Rex Warner (London: Penguin, 1972).

years. He will die fighting on his own land, and his main concern now is to not bring his friends down with him.

Every decision-making problem has a moral dimension, and the more important the decision, the larger this dimension. As Albert Camus said: "It is never by chance that one takes the dishonorable way."

6

Statistics

And Pharaoh said unto Joseph, In my dream, behold, I stood upon the bank of the river: And, behold, there came up out of the river seven kine, fat-fleshed and well-favored; and they fed in a meadow: and, behold, seven other kine came up after them, poor and very ill-favored and lean-fleshed, such as I never saw in all the land of Egypt for badness: And the lean and the ill-favored kine did eat up the first seven fat kine: And when they had eaten them up, it could not be known they had eaten them; but they were still ill-favored, as at the beginning. So I awoke. And I saw in my dream, and, behold seven ears came up in one stalk, full and good: And, behold, seven ears, withered, thin, and blasted with the east wind, sprung up after them: And the thin ears devoured the seven good ears: and I told this unto the magicians; but there was none that could declare it to me.

And Joseph said unto Pharaoh, The dream of Pharaoh is one: God hath showed Pharaoh what he is about to do. The seven good kine are seven years: and the seven good ears are seven years: the dream is one.

And the seven thin and ill-favored kine that came up after them are seven years; and the seven empty ears blasted with the east wind shall be seven years of famine. This is the thing which I have spoken unto Pharaoh: What God is about to do he showeth unto Pharaoh.

Behold, there comes seven years of great plenty throughout all the land of Egypt: And there shall arise after them seven years of famine; and all the plenty shall be forgotten in the land of Egypt; and the famine shall consume the land; And the plenty shall not be known in the land by reason of that famine following; for it shall be very grievous. And for that the dream was doubled unto Pharaoh twice; it is because the thing is established by God, and God will shortly bring it to pass. Now therefore let Pharaoh look out a man discreet and wise, and set him over the land of Egypt. Let Pharaoh do this, and let him appoint officers over the land, and take up the fifth part of the land of Egypt in the seven plenteous years. And let them gather all the food of those good years that come, and lay up corn under the land of Pharaoh, and let them keep food in the cities. And that food shall be for store to the land against the seven years of famine, which shall be in the land of Egypt; that the land perish not through the famine. (Gen. 41:17–36)

THE SCANDINAVIAN SOIL was harsh to its offspring, who soon learned to turn toward the sea, richer in resources and more promising. The world of the Viking was composed of fishing and pirating, expeditions of conquest and discovery. The leader was the one who brought his men to victory. The king's power was first and foremost a military power, taxes and plundering were the sources of his wealth. A bad season was mainly one during which navigation was impossible.

From the rich soil of Egypt, and perhaps Mesopotamia, two millennia before men from the North entered into history, come the founding myths of agriculture and the management techniques our governments still apply today. Certainly, Scandinavia had its share of famines. Snorri Sturluson tells of how Domaldi, one of the mythical kings, the first descendants of Odin, was sacrificed after three consecutive bad harvests:

Domaldi succeeded his father Visbur and ruled over his lands. In his days there was famine and starvation in Sweden. Then the Swedes made huge sacrifices in Uppsala. The first fall they sacrificed oxen, but the season did not improve for all that. A second fall they sacrificed humans, but the season remained the same or was even worse. In the third fall the Swedes came in great numbers to Uppsala at the time for the sacrifices. Then the chieftains held a council, and they agreed that the famine probably was due to Domaldi, their king, and that they should sacrifice him for better seasons, and that they should attack and kill him and redden the altars with his blood; and so they did.[1]

The king's responsibility is of a magical or ceremonial order here. We are in the troubled period when the memory of Odin is still alive, during which history is gradually separating from myth. The harvest is the visible result of a complex interaction between human society and the supernatural powers that control the fertility of the soil, in which the king plays an important role. Perhaps he is a victim of *seid,* since his mother-in-law had put a curse on him. Perhaps he is paying for having used *seid* against his father, King Visbur, and for having surprised him and burned him in his castle. The fact is that the sacrifice had the anticipated result, since Snorri specifies that the king's son, Domar, who reigned after him, lived a long life, during which the country enjoyed prosperity and peace.

In Joseph's story, we are in the presence of another kind of responsibility, of a bureaucratic and temporal nature. The Pharaoh is interested in what will happen in fourteen years; he wants to anticipate the troubles ahead, and has the material means to do so. He will institute a special 20% tax, paid in goods, on agricultural products, and warehouse the proceeds during the seven years of abundance. A specially created team of bureaucrats, directed by a powerful minister, will supervise the operation. Let us say, for instance, that the harvest will be 125% of the usual level during the seven plentiful years, and 75% of the usual level during the years of scarcity. The distribution proposed by Joseph leads to the consumption of 125 −

1. *Saga of the Ynglings,* chap. 15. From the *Heimskringla.*

25 = 100% of the normal level during the first period, and of 75 + 25 = 100% of the usual level during the second period. The operation worked, and was even a great success, at least for the Pharaoh, who obtained the grain free and resold it at a high price (Gen. 41: 56–57).

It is clear than an operation of this kind requires a large bureaucracy, well beyond what the Viking kings ever had at their disposal. One needs to take a census of the population, create a network of royal warehouses, store 20% of the harvest for seven years, keep it well preserved, do the bookkeeping, then resell the stock at a rhythm that will last seven years, which means having a relatively precise idea of the population's needs. Aside from the usual police, customs officers and tax collectors, this would also require an army of accountants and statisticians. The Pharaoh understands this and creates a special organization. These scribes perform economic calculations, analytical accounting and management, and use their technical competence to access political power. Joseph himself, who had a fine career as Zaphnath-Paaneah and the Pharaoh's minister, is a prime example.

Even in ordinary times, governing two kingdoms from one end of Egypt to the other meant managing future resources, which required a powerful administration, capable of devising long-term policies based on forecasts, and of imposing these policies in the present. The economy was essentially agricultural and the country lived according to the rhythm of the Nile's growing seasons. It was necessary to regularly evaluate the harvests from one end of the country to the other, to estimate whether they would be sufficient, whether there was enough to store, or whether, on the contrary, the reserves needed to be consumed. They had to keep exact records of the additions and depletions, and try to guess what the next harvest would be like in order to efficiently manage the current fluctuations.

Joseph simplifies the Pharaoh's problem by predicting in

no uncertain terms not only the next harvest, but the next fourteen harvests. As a result, the policy to follow is easy to figure out: economize during the seven plentiful years and redistribute the savings during the seven harsh years. But when you don't benefit from the direct assistance of the Almighty, the real problem is that you never know what tomorrow will bring. If the harvest is good this year, you don't know if it will be good or bad the next, and thus if it is appropriate to warehouse or consume today. And if two harvests in a row are good, it doesn't help matters: should you continue to warehouse for the bad times that will come one day, or give in without second thoughts to the joys of consumption?

In the absence of premonitions, the Pharaoh can fall back on a few tried and true methods. The first one, which is still in vogue today, is to increase one's territory by conquest and annexation. In so doing, he diversifies his risk, and thereby diminishes it. The other agricultural lands of the Fertile Crescent, Palestine, Libya, Syria, all the way to Mesopotamia, are not subject to the same risks as Egypt. The fate of these regions depends on the rainfall over the double chain of hills in Lebanon and Anti-Lebanon, and not on precipitation over the distant Ethiopian mountain chain that feeds into the Nile. Even farther away, in Mesopotamia, we encounter an irrigation system fed by the Tigris and the Euphrates, originating in the Cappadocia mountains. Three economies, each with its own random influences yet independent of one another; if it doesn't rain in Ethiopia, it doesn't mean it won't rain in Libya. An empire reaching from Egypt to Mesopotamia would suffer famine only if it didn't rain from central Africa to Asia Minor. Simultaneous droughts in such distant regions can indeed occur, but far less frequently than local droughts. If for example we estimate that once every seven years there will be no floods of the Nile, that once every seven years the Tigris and the Euphrates will be dry, and that these events are independent of one another, the probability that they will occur simultaneously is only one in forty-nine, which is to say that during his reign, the emperor may consider himself safe from such a catastrophe.

The other method, if you want to live in peace with your neighbor, consists of trusting to the weather. If indeed the next year is bad, that doesn't necessarily mean that the following year will be bad. There may indeed be climatic cycles, or manifestations of divine anger, but in the absence of convincing indicators it is reasonable to assume that the weather next year is independent of the weather this year. The consequence is simple. If we consider, as we did above, that the years of drought have a frequency of one in seven, the probability that there will be two consecutive droughts is only one in forty-nine, and the probability that there will be seven in a row is one in 823,543. This event is so improbable that its occurrence would be a convincing indicator of supernatural intervention, or of a mistake in the model.

At the heart of these arguments, as with all statistical analysis, is the notion of independence. It can be spatial, due to distance, or temporal, due to forgetfulness or the absence of memory. It is true that from a certain point of view, independence doesn't exist. In the final analysis, atmospheric circulation depends on the sun's rays and the earth's rotation. The local disturbances we observe here and there are only the consequences of complex interactions taking place around the globe. Precipitation in Ethiopia and in Cappadocia have common but distant causes. The astronaut sees this as he orbits the earth, taking in a hemisphere at a glance and admiring how cloud systems form and dissolve across oceans and continents. Their movements are sufficiently varied, their outcomes sufficiently unpredictable, for the proportion of rainy days in Ethiopia to be the same, whether we take into account the totality of the period or solely the rainy days in Cappadocia. In other words, the fact that it is raining in Addis-Ababa doesn't modify (or too slightly for us to notice) the probability that it will rain in Ankara, and this is what we mean when we say that these events are independent. Similarly, the fact that it is raining today in Cairo hardly tells us anything about what the weather will be like there in a year. In principle, the answer is contained in the state of the atmosphere today. But in three hundred sixty-five days, enough time will have passed to dilute the memory of the shower that has just fallen among so many other influences large and small—the flapping of a butterfly's wings or a cyclone over the China Sea. On the scale of one year, meteorology has no memory. The person who knows that it has rained today and the person who doesn't know have an equal chance of predicting rain a year from now.

In statistics we isolate events, which we conjecture to be random, in a very precise mathematical sense. We can imagine this by picturing for each event a large urn filled with red balls and green balls; each time there is a decision to be made, God pulls a ball out of the urn. If the ball is green, the event will occur; if it is red, it won't. The statistician's job is to guess the proportion of red balls to green.

In the simplest case, the balls are chosen independently. This means that if God draws several times, the result of the last draw is not influenced by the ones that came before it. One way of guaranteeing this independence is by putting the ball back in the urn and conscientiously shaking it after each draw to homogenize the distribution. The empirical frequency, that is, the number of times the green is chosen divided by the number of draws, should be close to the proportion of green balls in the urn, the estimate getting closer as the number of draws increases. Another way of creating independence is by drawing the balls from separate urns. The probability of drawing two green balls, one from each urn, is obtained by multiplying the probability of drawing one green ball from the first by the probability of drawing one green ball from the second. For example, if two events each have a probability of $1/2$ and are independent, the probability that they will occur together is $1/2 \times 1/2 = 1/4$.

There are several ways of creating dependence (generally called a correlation) between certain events. We can draw balls simultaneously from the same urn. For this, we must introduce two new colors, black and white. If we call X and Y the predicted events, the color code will be:

green = X and Y occur
white = X occurs, Y doesn't
black = Y occurs, X doesn't
red = neither X nor Y occurs

and as before, God will draw a ball whenever there is a decision to be made. If the four colors are in equal proportion, 25% each, we have the same probability assigned earlier for two independent events of a $1/2$ probability, and this draw would be the exact equivalent of a double draw in the two bicolor urns. We therefore can rightfully say that X and Y are independent. At the other extreme, we can leave out the white and black balls. In this case, X and Y are linked in the strictest sense, since we never observe one without the other; we don't even try to separate X and Y to see if one is the cause of the other, we simply note that they always appear together.

Between these two extremes lie all the degrees of correlation. As an example, let's examine what it means when the proportions are 30% green, 20% white, 20% black, and 30% red. If we are only interested in event X, we notice that it takes place when there is a green ball or a white ball, that is, that it has 30 + 20 = 50% chance of occurring. This is therefore the probability we assign X in the absence of all other information. But if we know in addition that event Y occurred, then the ball that was drawn must have been green or black. Given the relative proportions, there are three chances out of five that it will be a green ball, and therefore that X will occur as well. This supplementary information brings the probability of event X to 60%, instead of 50%. The fact that Y occurred increases the probability that X will occur: we say that events X and Y have a positive correlation.

We can of course take correlations into account, but the heart of statistics is the study of independent events. The most beautiful result, and the most universal one, is the "central limit" theorem, which describes with great precision the result of a large number of independent draws. Let's suppose, for example, that an urn contains white balls and black balls in equal proportions. We intuitively suspect that if we conduct enough draws, we should observe just about as many white balls as black balls, while realizing that our luck may be bad and we may draw only black balls. The "central limit" theorem gives us the relative frequencies of these odd cases, and allows us to recognize that they diminish rapidly with the number of draws. Thus, if we conduct 1,500 draws, we have an astronomical number—$2^{1,500}$—of possible scenarios, but for half of them the observed frequency of white balls will be between 49 and 51%, that is, it will be less than 1% away from the exact proportion. The probability of observing white balls between 49% and 51% of the time in 1,500 drawings is thus 0.5, and rises to 0.954 if we conduct 10,000 drawings, 0.999 if we conduct 27,000. This means that in the latter case we have fewer than one chance in a thousand of being wrong if we situate the exact proportion within 1% of the frequency observed.

The first lesson of the "central limit" theorem is that precision increases as the square root of the number of drawings. If we chose a given level of confidence, $^{1}/_{1000}$ for example (that is to say if we want to have less than a one in a thousand chance of being wrong), and if the exact proportion of the white balls in the urn is $^{1}/_{2}$, the exact frequency will be within a 0.33% interval of 50% in either direction after one hundred drawings, 0.033% after ten thousand drawings, and 0.0033% after one million drawings. The interval is reduced by a factor of 10 each time the number of drawings is multiplied by 100, the level of confidence remaining fixed at $^{1}/_{1,000}$. In other words, deviant drawings continue to exist—it is still possible to draw one hundred, ten thousand or one million white balls consecutively—but their proportion diminishes. We can interpret this result by saying that independent random errors around zero tend to balance out where we add them up.

The "central limit" theorem also has a geometric interpretation, which we obtain by representing the result of a large number N of independent draws from an urn containing an equal number of black and white balls. Let us count the white balls at the end of the N draws. We may wind up without any, if we only drew black balls, with N, if we only drew white balls, or with any of the cases in between. For each number n between 0 and N, we indicate the number of scenarios that resulted in this total number of white balls. We obtain a characteristic bell-shaped curve, called a Gauss curve, symmetrical around the mean value $N/2$ (if the urn contained as many white balls as black balls). The universality of this curve, omnipresent in all domains of science and technology, is one of the consequences of the "central limit" theorem.

Whenever we collect measurements a Gauss curve appears. The height of conscripts the day of registration, the rounding-off errors in calculations, the experimental measurements of physical constants are all distributed on a bell-shaped curve around their mean value. This is easily understandable if we consider that a man's height depends on constant factors that affect the entire population (type of food, the gene pool), but also on individual parameters (taste and life-style, physical activity, heredity, and mutations) which are distributed randomly. Each individual receives his parameters at birth, and as these allocations take place independently across the population, the distribution of heights in the population conforms to the "central limit" theorem. Similarly, each physical measurement is tainted with errors from various sources, due notably to the limits of the experimental apparatus and to the precision of the instruments used. Reproducing an experiment and taking measurements again is essentially a new drawing process, independent from the previous one, and thus falls within the range of the "central limit" theorem.

This theorem is so powerful that it extends to areas in which chance plays no apparent role. This is the case with arithmetic. Let's recall first that a number is prime if it has no divisors (except 1 and itself, of course). Thus, 2, 3, 5, 7, 11, 13, 17, 19, 23, 29, 31, 37, are prime; this is just the beginning of an infinite sequence, which has fascinated mathematicians since antiquity. If a number is not prime it breaks down into the product of prime numbers; 6 has two prime factors ($6 = 2 \times 3$); 210 has four ($210 = 2 \times 3 \times 5 \times 7$). To be divisible by 6, a number must be divisible by 2 and by 3; half of all numbers are divisible by 2, a third divisible by 3, and a sixth divisible by 6. Of course, $1/6 = 1/2 \times 1/3$. We can express this by saying that the probability that a number will be divisible by 6 is equal to the probability that it is divisible by 2, multiplied by the probability that it is divisible by 3, as if being divisible by 2 or 3 were independent random events. But we do not draw lots from an urn to decide whether a given number is odd or even. There is nothing random about a number being prime, or divisible by 2, or by 3, and interpreting the equality $1/6 = 1/2 \times 1/3$ in terms of probability theory is a strange idea at best, with no foundation in mathematical reality. Nevertheless, will we take it one step farther and apply the "central limit" theorem, as if we were truly dealing with chance? This would be highly paradoxical, since what could be more deterministic than the series of whole numbers.

And yet, in 1939 Marc Kac and Paul Erdös showed that the number of prime factors follows a Gaussian distribution. More precisely, the number of prime factors of an integer m is on the order of log log m, and the proportion of integers m for which this number is included between:

$$\log \log m + a\sqrt{2 \log \log m}$$

and:

$$\log \log m + b\sqrt{2 \log \log m},$$

is given by the surface situated under the Gauss curve $\pi^{-1/2}\exp(-x^2)$ between $x = a$ and $x = b$. It is as if the infinite series of integers were the result of a series of independent draws using prime numbers. As if God had first made the prime numbers, then constructed the others by drawing them at random. The first day he picked the 2 and the 3, and made 6. The second day he picked the 2, the 3, the 5 and the 7, and made 210.

All this is static. If now we want to add a dynamic dimension to the notion of independence, we come to Brownian motion. This was first seen in the erratic movement of microscopic pollen particles suspended in a liquid. Discovered in 1827 by the Irish botanist Robert Brown, identified in 1905 by Einstein and Smoluchowski as the effect of the random movement of the surrounding molecules hitting the pollen particles, its mathematical theory was created by Norbert Wiener and Paul Levy. It is today one of the main tools for creating models of phenomena dependent on time and chance; it is no less ubiquitous than Gaussian distribution, to which it is closely related.

Brownian movement is by definition a movement totally devoid of memory. A particle of pollen, moving in water, doesn't know when the next shock will occur, nor in which direction or how forcefully it will be propelled. The mathematical model idealizes this situation, and creates a particle that at each instant forgets where it came from, looks at where it is, and decides where it wants to go. In other words, the instantaneous displacement must be independent of past history; we can imagine it as being continuously chosen at random. Picture a trajectory which at each instant changes direction and speed, and yet which is continuous. What a maze it will chart on the graph! Whether we enlarge certain portions or try to look at it from a distance, we find the same structure at every level, a kind of broken line groping around our field of vision. Physically, we cannot get any smaller than the molecular scale, but mathematically we can enlarge it indefinitely, and we always find the same general appearance. On any time scale, short as it may be, the particle is constantly changing direction. This is why the trajectories of Brownian motion evoke the kinds of curves that are continuous but have no tangents which mathematicians of the last century considered somewhat dangerous curiosities. Physicist Jean Perrin made note of this in his book *The Atoms,* which attracted the attention of mathematician Norbert Wiener.

It was Wiener who, in 1923, provided the first rigorous mathematical definition of Brownian movement. The hard part was to show that a mathematical entity endowed with all the properties physicists commonly attributed to Brownian movement existed. Wiener resolved the problem by focusing on two fundamental properties. On the one hand, all trajectories must be continuous. On the other hand, once the position of the particle has been observed at instant $t = 0$ (and is therefore known), its position (which is random) at some other instant t must be governed by a Gaussian distribution, the parameters of which naturally depend on the time t that has elapsed. All other properties of Brownian movement can be deduced from these.

Once Brownian movement was established on solid mathematical ground, it began to play a central role in the creation of models for aleatory or random phenomena. When transmitting a radio signal, for example, there are always various kinds of electronic interference; the difficulty lies in filtering the signal, that is, in separating it from background noise. The first filter was built by Wiener himself, who thereby inaugurated his newly forged tools. For a long time his results were kept a military secret; they were being applied to a new and extremely sensitive technology—radar. Today, more efficacious filtering methods have been discovered and are used on the automatic pilot and inertial guidance systems of commercial airlines or submarines. But the applications of Brownian motion go well beyond the treatment of signals. Whether one studies the spread of an epidemic or the diffusion of heat, Brownian motion is the basic tool for creating a model.

In recent years, Brownian motion found another application. At the end of the last century, a mathematician named J. Bachelier defended a thesis in which he proposed, by using Brownian motion, to create a model for the price of stocks on the Stock Exchange.[2] Neither the stock markets nor mathematical techniques were sufficiently advanced for this idea, and Bachelier faded into bitterness and obscurity. It wasn't until Wiener came along that Bachelier's work experienced a revival, which turned to triumph beginning in 1973, when Fischer Black and Myron Scholes demonstrated their famous formula for the evaluation of stock options relying on stochastic calculus, the modern outgrowth of Brownian motion.

We should first make it clear that there is nothing arbitrary about the idea of creating a model for stocks based on Brownian motion; instead, it reflects specific ideas about the behavior of stock markets. If we accept the idea that they are efficient, that is, that the price of a stock reflects all the information available on that stock, then we accept the notion that variations in this price reflect the arrival of new information. Some of this information was predictable, some of it wasn't. If the market has done its work, the former is already accounted for in the price of stocks: a rise or fall has already taken place based on this predictable information, and when the anticipated change occurs it no longer affects the price. The price therefore depends only on the truly new segment of the infor-

2. "Théorie de la spéculation," *Annales scientifiques de l'École normale supérieure* 17 (1900): 21–86.

mation, the part that is unpredictable based on available information. From this point of view, it is natural to model the market's official prices on a process with independent increments, that is, in the final analysis, on Brownian motion.

If we look at it from another perspective, brokers create prices, and it is hard to believe that these thousands of people sitting in front of computer screens all over the world, each with his own intuitions and fears, go to all this trouble just to construct a Brownian motion. The truth, no doubt, is that this hypothesis suffices to explain (if not to predict) most price changes in most stock market situations, but that it is in the remainder that human ingenuity comes into play and that fortunes are made or lost. In addition, the use of Brownian motion in financial theory goes beyond the simple adjustment of prices, allowing us to determine the price of certain financial products, such as options.

An option is a contract by which the seller commits himself to sell a certain share within three months at a price determined today. The buyer has the right not to exercise the option if this price turns out to be higher than the price at the date the option falls due. The holder of a large number of shares who wants to protect himself against a fall in price might buy the corresponding options with the idea that he won't use them if prices rise or remain stable. The person who sells the option takes a risk, and has to be compensated. The question of what a fair price for an option is was resolved by Black and Scholes in 1973. Their formula doesn't address the probability of a rise or fall in price but only what we call the stock's volatility, that is, its tendency to oscillate. As a result, we don't ask the brokers to make predictions, even statistical ones, as to the evolution of the stock's price, but to agree on the level of its volatility, that is, to identify one of the parameters that govern a Brownian motion. As such, you don't have to know if the stock will go up or down in order to determine the price of the option. This remarkable result is the basis for all modern theories of finance, and popularized Brownian motion among people who didn't consider themselves mathematicians.

There is a beautiful simplicity to reducing statistics to the "central limit" theorem and its various metamorphoses, including Brownian motion. Unfortunately, the statistician is confronted with other problems. More than any other scientist, he suffers the painful experience of being unable to validate a model: the only response he can provide with certainty is a negative one.

We started with urns containing certain proportions of balls in various colors, and we asked ourselves if the frequencies of these colors observed in the course of a succession of independent draws varied from the real proportions. Unfortunately, the statistician is not given this privileged information. All he has is a series of observations—with any luck a large series. He cannot identify chance, recognize that he is dealing with balls in an urn, guess the urn's contents, and state that these observations result from independent draws. The possibility still exists that other mechanisms are at play that we are mistakenly referring to as chance, just as it is not a matter of chance when a cheater shuffles and cuts the cards in such a way as to distribute the hand he desires. The statistician can never confirm a probabilistic model. At best he can invalidate it based on the incompatibility of the observations with the model proposed, that is, he can say that this model cannot possibly be the right one, otherwise the probability of the actual observations would be so small that they would never have occurred.

The statistician takes a series of observations and checks whether they are compatible with a given probabilistic model. To do so, he computes from the given model the theoretical probability of the observations in questions. If he finds it to be too low, say 1/100 or less, he will conclude that the model is not compatible with the observations. He thereby follows the old heuristic principle according to which events with low probability don't occur, a principle that in this case leads to dismissing the model proposed. On the other hand, if the model is judged to be compatible, that is, if the probability of the series observed using this model is on the order of $1/10$ or more, we still cannot conclude that the model is valid, even if the theoretical probability of the observations is $9/10$, $99/100$, or better. This is because we can never rule out the possibility that another, better-chosen model would have had even better results, or even that in the final analysis the phenomenon considered does not depend on chance after all.

This of course is the problem with science in general. We haven't been admitted into the Creator's workshop, and if we had, as many good authors have pointed out, we might have suggested some alternate solutions. All we can do is draw up blueprints and check whether things happen as if these blueprints were correct. The scientist is constantly looking for the decisive experiment that will allow him to invalidate a theory. But in statistics, this process of verification extends to the level of technology. In factories and government agencies, in the field and at universities, the statistician creates hypotheses and tests them. Quality control, for example, is a matter of verifying whether the proportion of defective pieces that come off an assembly line is below a given threshold. The statistician will assume that this is the case, will fix a level of confidence, and will test the assumption on random samples. He will intervene if the test rejects the hypothesis, that is, if at the level of confidence fixed the hypothesis turns out to be incompatible with the results obtained.

Though unambitious, this manner of proceeding by invalidating models is extremely general and versatile. For quality control, it is the rejection of the model that is important, because it shows that the values attributed to certain parameters are not correct, that the theoretical thresholds are not respected. In other situations, the compatibility of a probabilistic model with the observations will be important, since it will signify that we cannot exclude the hypothesis that the phenomenon being studied is based on chance. Whether it is a matter of grading multiple choice questions, of testing a new medication, or of analyzing psychology experiments, the question must always be asked: would chance alone have done equally well? If I am given one hundred questions, each with four possible answers, on a subject I know nothing about, and if I choose my answers randomly, I can hope to get twenty-five of them correct. This is why, for this type of test, twenty-five or fewer correct responses does not raise your score above zero. Grading begins above this point, which doesn't mean this method won't sometimes be unfair.

Only if we manage to exclude chance can we advance other explanations—the student's knowledge, the benefits of a certain therapy, or telepathy. As long as the statistical tests don't exclude it, nothing can be proven. This seems elementary, but it isn't. Few people realize that by responding randomly you will often get an answer right, and on occasion you will also score spectacular strings of success.

Note that the statistician, however, can never prove the presence of chance. He can exclude it in certain cases, or leave the possibility open, without being able to guarantee that it will be present; the fact that this explanation cannot be excluded is enough. The only case in which the statistician can state that chance is present is when he introduces it himself. In conducting a poll, for example, he will be careful to select his sampling randomly, that is, one way or another, to organize independent selections of balls from an urn. This seems simple, and if we are talking about examining products at the end of an assembly line, we can indeed proceed in this way. But it is another story if we want to poll a population on its political opinions or sexual habits. How should we question people? On the street? What about people who never go out, or who drive everywhere? By telephone? Not everyone has one, and several people may share the same number. At home? How can you randomly choose from an entire territory, rural and urban areas combined? What should we do if someone refuses to answer? Ask someone else the same question, or account for this person otherwise? And what about the answers obtained? Who can guarantee their honesty? People don't readily admit to certain political preferences, which nonetheless show up in the polling booth. Should we take certain biases into account in the answers? How? The statistician soon learns to establish extremely rigorous protocols for his experiments, in which chance plays but a minor role.

The great discovery of these last years, in fact, is that statistics can function perfectly well without chance. The spread of computer techniques in management has led to the accumulation of enormous masses of data in all areas, and their simple classification, not to mention their interpretation, poses considerable problems. Traditional statistical methods such as factorial analysis are available to do this, but new methods of automatic classification and of data analysis have been developed which still call themselves statistics but do not rely on probabilistic models. We recognize shapes by representing the data as points in a space with many dimensions, for example, and by seeking to separate them into masses that are as distinct as possible. This is work that is done with the naked eye in two dimensions, but which requires the assistance of a computer and of complex calculations if more than three parameters are being treated. This is a geometry problem that no longer has to do with randomness. But the situation may reverse itself again. More and more, the word "statistics" signifies "automatic treatment of data," and the development of the subject is necessitated by larger and larger quantities of computerized data. In addition to the traditional problems of classification and interpretation new ones are arising. For instance, the data need to be compressed: computer memories are quickly becoming saturated with the masses of information to be stored, and the time required to access data is becoming prohibitive. It is therefore becoming necessary to treat incoming information so that it occupies the fewest possible memory bits; in this process of breaking down and rebuilding data, classic probabilistic and statistical models are paradoxically regaining ground. None of this would be possible without the considerable progress of the last ten or twenty years in computer technology. But technological progress by itself would not have been enough to deal with the explosion of data; it was necessary to develop effective methods of computation which take advantage of the internal structure of the computer. We are now beginning to design computers according to the kinds of calculations they must perform. Like architects who deal with the flow of travelers at a train station or airport, computer designers are working on the circulation of bits—of 0s and 1s—through the computer at various stages in the calculation. They could not do this without probabilistic models and classical statistics, which are thereby making a comeback.

Statistics is not designed to prove the existence of chance, or to detect its presence. On the contrary, it is based on an initial premise, which is that the world is probable. Like all of us, the statistician begins with the principle that the world exists, but asks something more: that it be probable. It is theoretically possible, in any case perfectly compatible with probability theory, that starting tomorrow whenever someone flips a coin it will land face up, and that the roulette wheel's ball will land on only red, or even, or zero. While such a situation is infinitely unlikely, it is possible; if it were to come to pass, none of the statistics textbooks would have to be changed. They would teach us that if the Creator were to start the universe over, he would probably fabricate one that behaved more normally.

But in the meantime, we would be forced to live in an improbable universe, in which rivers return to their sources and entropy decreases over time.

We speculate that this is not the case. We believe we live in a universe in which events that have too slight a probability don't happen, and we conduct ourselves accordingly. Until now, experience hasn't proven us wrong, but who knows what the future may bring.

Conclusion

During the aforesaid night, when King Olaf lay in the midst of his troops, he stayed awake for a long time, praying to God for himself and his men, and slept but a short while. Toward morning sleep overcame him, and when he awoke, day broke. The king thought it was rather early to wake the army. Then he asked where the skald Thormoth was. He happened to be near and asked what the king wanted of him. The king said, "I would have you recite some lay for us." Thormoth arose and spoke in a very loud voice so that all the army could hear him. He chanted the "Old Lay of Bjarki," of which this is the beginning:

Day has come,
the cock shakes his wings.
'tis time for thralls to take to their tasks.
[Awake, ye friends,
be wakeful ever,
all ye best men
in Athils' court.

Har the hard-gripping,
Hrolf the bowman,
men of noble line
who never flee:]
I wake you not to wine
nor to women's converse,

but rather to Hild's[1]
hard game of war.[2]

THE BATTLE TOOK PLACE that day, the 29th of July, 1030, at Stiklestad, and King Olaf perished in it along with the better part of his army. His true destiny began after his death. The people and the barons who had banded together to crush him soon repented for having thoughtlessly left Norway in the hands of the King of Denmark. With great pomp they transferred the body of Olaf Haraldsson to the Cathedral of Nidaros, and retrieved his son Magnus from his Russian exile in Novgorod in order to crown him. Olaf's reputation for holiness quickly spread through all the Nordic countries, and his tomb became one of the most popular pilgrimage sites of the Middle Ages. The cathedral was largely destroyed during the Reformation, and today no one knows where the remains of Saint Olaf lie.

Tormod the skald fought under King Olaf's flag and was seriously wounded by an arrow that struck him in his side. He died that evening as he was pulling the arrow that had struck him from his own body. Snorri Sturluson recounts that the barbs of the point were covered with heart fibers, red and white, and that Tormod said, "Well has the king fed us. I am fat still about the roots of my heart."[3]

Men die, only art survives. The themes of the old song, daily labor and hand-to-hand combat, the heavy habits of the present and the worried hope for the future, still resonate today. Above the melee, beyond the noise and fury, a call for earthly harmony rises. It is because they hear this call that Olaf Haraldsson's men are willing to follow him in hopeless battle.

And what about me, why am I willing to devote my life

1. A Valkyrie: Hild's game is war.
2. *Saint Olaf's Saga,* chap. 208, in the *Heimskringla.*
3. Ibid. chap. 234.

to science? Is it to find myself tossed about by chance, incapable of making predictions, reduced to recording what exists? Why should I engage in this battle, following in the footsteps of so many others, if it inevitably leads to declaring chance king of the universe?

It is because I too have heard Tormod's song. There is more to science than just chance, even though I devoted this book to the topic. In my work as a researcher, I explore other ideas in which chance plays no part. Geometry, general relativity, the dynamics of conservative systems, the physics of elementary particles, are all theories of almost superhuman beauty, in which I find the same harmony expressing itself in the same mathematical form, that of a variational principle.

A variational principle is a mathematical criterion that allows us to distinguish a particular solution from among a multitude of possible solutions. The simplest and best known is the one that says that the straight line is the shortest distance between two points. To put this variational principle into practice, we first need to define the distance between two points and the length of a curve. Then, among the infinite variety of curves traveling between two points, we have to find the one that is shortest in length. This is how we define the straight segment, and all its other characteristics follow from this fundamental property. Suddenly we have constructed Euclidian geometry, not based on Euclid's axioms, but on one simple property.

This undertaking would be of interest only to mathematicians if it weren't also at the heart of physics. As early as the seventeenth century, Pierre de Fermat stated that luminous rays minimize optical length. This meant that all geometrical optics (at the time, the most advanced branch of science), were subject to a variational principle. To take advantage of this, one must first define optical length, which is not exactly the usual (geometric) length, but which is related to it by the refraction index of the area it traverses. Among all the possible trajectories, we then have to seek the one with the shortest optical length. Mysteriously, this is the one that the light ray follows. We can calculate it: it will be a straight segment if the refraction index in that area is constant, a more complicated curve if it varies. All the laws of geometrical optics, whether they concern reflection, refraction, or lenses, follow from this single principle.

It was again Fermat who reduced the laws of mechanics to one simple variational principle, which he called the principle of least action. The latter survived two revolutions of contemporary physics unscathed, adapting both to general relativity and to quantum mechanics. It is still at the heart of our body of knowledge; each new upheaval leads to a reorganization around it, but reinforces its central position.

As early as the seventeenth century, men marveled that such deep results in physics could find such a simple mathematical formulation. Why does this principle of least action play such a role? The question reverberated for the philosophers of the day, notably for Malebranche and especially for Leibniz, whose work is the most daring attempt yet to respond to this question. To say that this world is the best of all possible worlds is a formulation too easily ridiculed, but it translates the experience and enthusiasm of the period. As Leibniz was discovering the language of modern mathematics, he watched the physical world gravitate to it as to a mother tongue. Variational principles express themselves naturally in this language, and we saw the newborn methods of calculus translate them into practical computations. Leibniz sensed that science would develop around this central organizer, and wanted to understand the reasons before seeing the product. We can only admire his daring—and share his admiration.

More than three centuries later, we are seeking the same synthesis, infused with the same vision. Since Plato's *Parmenides,* we know that truth does not allow itself to be grasped, that if there is an ultimate reality it retreats the closer we come to it, finally vanishing into insignificance. From elementary particle to elementary particle, from psychological analysis to psychological analysis, the descent is subtle and endless. This path can only lead to the discovery of contingency; and so it is that chance will be our constant companion. But we are looking for a different path, an ascent that will bring things together rather than disperse them, a path on which we will part with chance, just as Dante parts from Virgil at the entrance to Paradise. Then beauty will be our guide.

Index